Environmental Studies:

A Practitioner's Approach

AF818601

Environmental Studies:

A Practitioner's Approach

Prof. SOLI J. ARCEIVALA
Formerly, Chief Environmental Health, UN/WHO SE Asia Region and Director, National Environmental Engineering Research Institute, Nagpur

Prof. SHYAM R. ASOLEKAR
Centre for Environmental Science and Engineering
Indian Institute of Technology Bombay, Mumbai

Tata McGraw Hill Education Private Limited
NEW DELHI

McGraw-Hill Offices

New Delhi New York St Louis San Francisco Auckland Bogotá Caracas
Kuala Lumpur Lisbon London Madrid Mexico City Milan Montreal
San Juan Santiago Singapore Sydney Tokyo Toronto

Tata McGraw-Hill

Published by Tata McGraw Hill Education Private Limited,
7 West Patel Nagar, New Delhi 110 008

Copyright © 2012, by Tata McGraw Hill Education Private Limited

No part of this publication may be reproduced or distributed in any form or by any means, electronic, mechanical, photocopying, recording, or otherwise or stored in a database or retrieval system without the prior written permission of the publishers. The program listings (if any) may be entered, stored and executed in a computer system, but they may not be reproduced for publication.

This edition can be exported from India only by the publishers,
Tata McGraw Hill Education Private Limited.

ISBN (13): 978-1-25-906389-3
ISBN (10): 1-25-906389-5

Vice President and Managing Director—Asia-Pacific Region: *Ajay Shukla*
Publishing Manager—Professional: *Praveen Tiwari*
Editorial Researcher—Science, Technology and Computing: *Sushil Gupta*
Sr. Copy Editor: *Neha Sharma*
Sr. Production Executive: *Rita Sarkar*
Manager—Sales and Marketing: *S Girish*
Deputy Marketing Manager—Science, Technology and Computing: *Rekha Dhyani*
General Manager—Production: *Rajender P Ghansela*
Manager—Production: *Reji Kumar*

Information contained in this work has been obtained by Tata McGraw Hill, from sources believed to be reliable. However, neither Tata McGraw Hill nor its authors guarantee the accuracy or completeness of any information published herein, and neither Tata McGraw Hill nor its authors shall be responsible for any errors, omissions, or damages arising out of use of this information. This work is published with the understanding that Tata McGraw Hill and its authors are supplying information but are not attempting to render engineering or other professional services. If such services are required, the assistance of an appropriate professional should be sought.

Typeset at Text-o-Graphics, B-1/56 Arawali Apartment, Sector 34, Noida 201 301 and printed at

Cover Printer:

Cover Design: Kapil Gupta

Cover Photos Courtesy: S Balasubramanyam

RXCYCRAHDXZQB

The McGraw-Hill Companies

"Sun the Source of All Energy in the Ecosystem"

A painting done in 2008 in acrylic on canvas by Prof Soli J Arceivala presently displayed at the National Environmental Engineering Research Institute (NEERI), Nagpur

It is important for all of us to remember the "fundamental duty" clause and the "directive principle" articulated in our *Constitution*:

"It shall be the duty of every citizen of India
to protect and improve the natural environment
including forests, lakes, rivers and wild life,
and to have compassion for living creatures."
Article 51A(g), the *Constitution of India*

and

"The State shall endeavour to protect and improve
the environment and to safeguard
the forests and wild life of the country."
Article 48A, the *Constitution of India*

in a professional manner, the general philosophy behind environmental protection. This philosophy is timeless.

Among the many people to whom we owe thanks, we must mention Prof Narayan Dhabolkar, the then Director, VJT Institute, Mumbai, and Prof Devang Khakhar, Director, IIT Bombay, for their encouragement and support. We would also like to thank Prof Prashant Bhave of VJT Institute, Mumbai, for his contribution in preparing Chapters 5 and 9. Last, but not the least, we must thank our own families for their patience, encouragement and support all along.

Soli J Arceivala
Shyam R Asolekar

Contents

Preface *vii*
Introduction *xv*

1. Some Essentials We Must Know First **1**

1.1 Global Ecosystems and Fate of Pollutants *2*
1.2 Policies and Strategies for Protecting the Ecosystem *18*
1.3 Environmental Quality Assessment—Use of Criteria and Standards *22*
1.4 Natural Resources and Need for Conservation *27*
1.5 The Philosophy of Waste Disposal—Some Indian Standards and Environmental Laws *34*
1.6 The Role of Statistics *38*
References *41*
Exercise Questions *41*
Project/Assignment Topics *42*

2. Environmental Aspects of Energy Production **43**

2.1 Emissions in Traditional Power Generation with Fossil Fuels *44*
2.2 Emissions from Industry, Commerce and Services *47*
2.3 Emissions from Livestock *48*
2.4 Emissions from the Transport Sector *48*
2.5 Important Alternative Sources of Energy *52*
2.6 Alternative Sources Already in Use *53*
2.7 Alternative Sources of Energy Ready to Use *58*

2.8 Alternative Sources of Energy Needing Development *73*
References 76
Exercise Questions 76
Project/Assignment Topics 77

3. Rainwater Harvesting in Urban and Rural Areas 79

3.1 Rainwater Harvesting in Urban Areas, and its Hazards *80*
3.2 Rainwater Harvesting in Rural Areas, and its Advantages *89*
3.3 Some Common Aspects to Urban and Rural Supplies *97*
3.4 Summary of Recommendations *102*
References 104
Exercise Questions 104
Project/Assignment Topics 104

4. Preventive Environmental Management 107

4.1 The New Management Challenges *108*
4.2 Modern Trends in Environmental Management *112*
4.3 Some Issues in Town-Planning and Infrastructure Development *116*
4.4 Environmental Impact Assessment *119*
4.5 Case Study: The EIA of a Water Resources Development Project *122*
4.6 Life Cycle Assessment *125*
4.7 Case Studies of LCA Applications *128*
References 128
Exercise Questions 129
Project/Assignment Topics 129

5. Air Quality—Ambient and Indoor 131

5.1 Earth's Atmosphere, Some Air Pollutants and their Health Effects *132*
5.2 Some Concepts from Meteorology Affecting Air Movements *142*
5.3 Industrial Stack Emissions and Control Strategies *145*
5.4 Traffic Pollution and its Control *147*
5.5 Indoor Air Pollution and Ventilation *150*
5.6 Ozone Depletion and Acid Rain *152*

5.7 Noise Pollution *154*
Exercise Questions *155*
Project/Assignment Topics *156*

6. Water Resources and Supply Systems **157**

6.1 Principles of Water Treatment and Supply *158*
6.2 Drinking Water Quality for General Consumption *162*
6.3 Water Quality for Industries and Boilers *166*
6.4 Development of Water Resources—The Indian Scenario *167*
6.5 Conclusion *176*
References *177*
Exercise Questions *177*
Project/Assignment Topics *177*

7. Wastewater Treatment for Pollution Control **179**

7.1 Some Concepts in Wastewater Management *180*
7.2 Objectives of Treatment *182*
7.3 Wastewater Treatment Technologies, Factors in Choice *184*
7.4 Resource Recovery, Waste-to-Energy *189*
7.5 Some Treatment Methods for Domestic Wastewater *190*
7.6 Will Total Sanitation Ever Come in Our Lifetime? *192*
7.7 Industrial Waste Treatment and "Zero" Discharge from Industries *194*
7.8 Thermal Pollution *196*
7.9 Marine Pollution, Coastal Waters (CRZ), Mangroves and Wetlands *197*
7.10 Coastal Waters and Coastal Regulation Zone (CRZ) *200*
7.11 Common Effluent Treatment Plants (CETPs) *203*
References *206*
Illustrative Example *206*
Project/Assignment Topics *208*

8. Wastewater Reuse for Irrigation, Industry and Community Uses **211**

8.1 Reuse of Wastewater in India *212*
8.2 Reuse in Irrigation, Soil-Crop-Water Balance, Health Hazards *213*

Introduction

As stated in the preface, the topics discussed in this book have been grouped from a pollution and infrastructure standpoint to sensitize young engineers to the dangers of abusing or neglecting the environment.

Chapter 1 explains a few essentials concerning the natural ecosystem and how our well-being is intricately affected by various actions and activities occurring elsewhere. Unless we know our ecosystem thoroughly and understand its complex webs and chains of responses, we know nothing of the environment. Unless our knowledge of the ecosystem and the fate of pollutants is sound, and our reasoning is clear, we would not be able to shape our environmental protection policies and strategies correctly. We also need to understand the difference between pollution control 'criteria' and pollution control 'standards' and how they are formulated since we have to maintain our health and well-being in the midst of so many different and new pollutants to which we are likely to be exposed.

The chapter also discusses a few important natural resources like water, land, forests, minerals, etc, which are involved in development, and have to be used carefully and conserved to match their natural rate of replenishment which is relatively slow.

Chapter 2 deals with energy production and its environmental and resource aspects. Every industry uses much electricity and the time has come to use it more efficiently and find alternative sources of energy that will help us reduce our carbon footprint and conserve our resources.

Chapter 3 points out the hazards of rainwater harvesting in urban areas while at the same time it points out that India will be transformed into a more prosperous country when harvesting is practiced more widely in rural areas.

Chapter 4 deals with the very important subject of Preventive Environmental Management beginning with environmental impact assessment, which every engineer, whatever his specialization, every businessman, every builder and planner must understand and act upon. It is a crucial chapter which no one should take for granted.

Chapter 5 deals with the quality of air we breathe both indoors and outdoors. Industries and traffic can be big air polluters.

Chapters 6 deals with water supply and its quality for industries and cities, and tells us that where there is water, there will be wastewater.

Chapters 7 and **8** deal with the complexities of affordable wastewater treatment and reuse in industries, irrigation and communities. From liquid wastes we pass to municipal solid wastes in **Chapter 9** and to the more difficult and hazardous wastes problems in **Chapter 10**.

We also see some useful activities emerging through corporate and community initiatives in **Chapter 11** including initiatives to the newest environmental problem of carbon emissions and global warming. This chapter also deals with the important topic of sustainability in everything we do.

Chapter 12 lists the vast opportunities that lie for consulting, contracting and other work in the environmental field. The chapter also covers cost of treatment and funding sources and methods.

Today we live in a world where in every two minutes:

- The world population increases by 800 people.
- Over 120,000 tonne of Greenhouse gases are emitted into the sky.
- Nearly ½ square mile of forest is destroyed in the world.
- More than 100 children (under age 5) die of easily preventable diseases.

It is evident that the world cannot be used on a "first come, first served" basis! We have to leave the world in a good enough condition for posterity.

As engineers, our twin objectives have to strengthen the economy and improve the environment at the same time.

Recommended Supplementary Activities

It is suggested that to improve the interactive aspect of the subject, site visits be organized for the students to supplement classroom work and the students be encouraged to develop a lifelong habit to use the library and internet to seek new knowledge.

Suggested site visits

- a water intake and water treatment plant,
- a wastewater treatment plant, OR a common effluent treatment plant (CETP),
- a sewage farm,
- a municipal solid waste disposal plant,
- a hazardous waste disposal plant,
- a thermal or hydro power generating station,
- a certified Green Building,
- a rainwater harvesting site,
- an automobile exhaust measurement station (PUC),
- a ventilated chlorination room,
- a manufacturing industry.

The library and internet

As new knowledge develops very fast in this day and age, everyone has to pick up the library and internet habit to keep abreast of developments. Hence, at the end of every chapter we have given "References" and "Recommended Additional Reading" which we hope will be followed.

Similarly, we hope the teacher will use the "Illustrative Examples" and "Assignment Topics" indicated by us at the end of every chapter to help the students better understand the text, and to develop the library and internet habit. Finally, to generate some lively discussion in the classroom some 'soft' and some 'hard' options have been indicated after each chapter which can be discussed with the students to broaden their understanding.

It was the great Indian sage, Swami Vivekananda, who perhaps unwittingly set the agenda for all environmentalists as early as 1905 when he said:

"To protect is the voice of the present; to prevent is the divine whisper of the future".

1.1 GLOBAL ECOSYSTEMS AND FATE OF POLLUTANTS

1.1.1 Health and Well-being

We have come a long way from the early 1850s when bacterial infections were predominant and gastro-intestinal epidemics (typhoid, cholera) occurred killing hundreds of people in the infected areas. It was soon recognized that such epidemics occurred in areas where sanitation was lacking. The areas had developed rapidly owing to new industries springing up there, but sanitation and water quality control had not yet followed up. It took some years to redress this situation in the western countries, although even today with all our technological advances we still have these dreaded epidemics in our country.

However, as gastro-intestinal infections diminished, other more virulent ones came up. As life became more complex, a variety of "poisons" found their way into public water bodies, among them are:

- Substances leached from minerals, metallic deposits and mining operations.
- Heavy metals dissolved from structures, metallic pipes, plastics, or added as water treatment chemicals.
- Polluting compounds from a variety of untreated industrial wastes.
- Poisonous compounds generated by natural algal growths.
- Pesticides washed down from agricultural fields into rivers.

Except agricultural chemicals which come from wide areas, the others have been increasingly diminished through process changes, chemical control, operational developments and various other methods collectively called "Cleaner Technologies".

In recent times, other effects like acid rain, radioactive fallout, traffic emissions, ozone hole, global warming and climate change, have additionally come to our attention. They are all global and difficult problems to solve.

The reason we want to study the ecosystems from the pollution viewpoint is to be able to find smarter ways of protecting our environment under these difficult circumstances as our health and well-being depend upon it. We find that every facet of man's activity affects the environment and man's health.

A short list is given below to show the wide variety of problems we face:

- *Drinking Water Quality:* Diarrhoea, dysentery, typhoid.
- *Other water-borne infections:* Filaria, malaria, dengue, etc.
- *Air-borne infections:* Various microbial and viral infections like colds, coughs, tuberculosis, etc.
- *Food-borne infections:* Microbial and viral contamination of food as well as problems caused by persistent substances in the food chain.
- *Traffic pollution (diesel and gasoline fumes):* Asthma and other allergies, cancer.
- *Industrial air pollution:* Asthma, allergies, pulmonary infections, cancer, etc.
- *Global warming:* Various direct and indirect health problems caused by climate change and sea level increase.
- *Ozone Hole (Radiation):* Skin cancer.

This list is by no means exhaustive, but only illustrative. Many more problems are caused by various other activities of a commercial nature which also affect our health and/or well-being indirectly, such as a few listed below:

- Heavy algal, mangrove and aquatic vegetation growths in lakes, rivers, estuaries and coastal waters caused by what is called "eutrophication" (excessive presence of nutrients in water).
- Damage to fisheries in coastal waters, lakes and rivers (either no fish or contaminated fish).
- Odours and smells affecting human well-being and land use values adjacent to water bodies.
- Ground water problems affecting agriculture, habitation and well-being.
- Health and other problems caused by climate change, radiation, etc.

1.1.2 Plants and Organisms

Before we see more of pollution-related ecosystems, we must mention some of the living plants and animals which constitute an important part of the ecosystem. In fact, all organisms found in nature are of concern in the management of environment. A partial list is given below:

Animal Kingdom

Vertebrates (humans, animals, fish, amphibians)
Invertebrates (molluscs, arthropods, worms, rotifers, etc.)

Plant Kingdom

Spermophyta (seed-bearing)
Bryophyta (mosses)
Pteridophyta (ferns)

Protista (Primitive)

Higher (algae, protozoa, fungi, slime molds)
Lower (blue-green algae, bacteria)

Bacteria

These are small, unicellular organisms with cell walls. They generally vary in diameter from 0.5–1.5 microns. More than any other group of organisms, the saprophytic bacteria are responsible for decomposition whereby waste organic matter is biochemically degraded and eventually mineralized or stabilized as explained further below (see Decomposers in Fig. 1.1).

Viruses

These are a specific group of organisms so small in size (nano?) that they cannot carry out any metabolic functions themselves and are therefore essentially parasites living on larger organisms. The only function they can carry out is to replicate themselves! (e.g., polio and infectious hepatitis viruses).

1.1.3 Global Ecosystems

The environment has two major components: *Living and Non-living*. A few of the living organisms were listed above. The non-living components are water, air, solar energy, etc. The global ecosystem consists of a set of inter-dependent actions and reactions between the living and non-living components as shown in Fig. 1.1. Ecology embraces physics, chemistry, biology, microbiology, and all the other sciences all in one go. These interactions go on continuously as described below because the *sun* supplies the required "energy" to keep things going. Life on earth depends on a steady input of energy from the sun. The whole world can be regarded as an ecosystem. All its individual units like lakes, rivers, soils, even a puddle of water can be cited as good examples of small ecosystems or eco-subsystems. This explains the multi-disciplinary nature of environmental studies. It includes biology, chemistry, engineering, everything!

Let us take *photosynthesis* as the starting point in the diagram (Fig. 1.1). Photosynthesis takes place in nature in the presence of solar energy, water and nutrients (carbon, nitrogen, phosphorus, etc.) and generates the basic food (green matter) on which all life depends. The three ingredients (solar energy, water and nutrients) are essential for photosynthesis to occur. Photosynthesis occurs in water of ponds, lakes, rivers, seas, etc., producing algae and other green matter, or on land in the form of trees, grass, forests, etc. Basically, photosynthesis occurs as follows:

Solar energy + water (H_2O) + nutrients and micro-nutrients (C, N, S, P, etc.) = Primary Production of algae, grass, trees and other green matter, as shown in the Fig 1.1.

With a little bit of chemistry, the following equation shows how CO_2 reacts with water and nutrients to give algae and produces oxygen at the same time.

$$106CO_2 + 122H_2O + 16NO_3 + HPO_4 + 18H \rightarrow \underset{\substack{\text{(algae)} \\ (2427)}}{C_{106}H_{263}3O_{110}N_{16}P_1} + \underset{\substack{\text{(oxygen)} \\ (3800)}}{138O_2}$$

From the equation we find that for every 2427 grams of algae, 3,800 grams of O_2 are produced. On this basis, the weight of oxygen produced is about 1.6 times the weight of algae produced by photosynthesis. Algae is the primary source of all food on the planet and together with the oxygen, it keeps the world functioning.

The oxygen helps to support life. All living things, human beings, animals, trees, etc., breathe in oxygen and give off CO_2. During daylight hours, trees undergo both respiration and photosynthesis. About 14% of the oxygen produced through photosynthesis is consumed back in respiration. The rest helps all living things to survive.

There are three different types of micro-organisms in the category called *decomposers*:

➢ *Aerobic* micro-organisms (which flourish when free oxygen is available)
➢ *Facultative* micro-organisms (which can exist when free oxygen is available or in its absence they breakdown nitrates (NO_3) and sulphates (SO_4) to obtain their requirement of oxygen, and
➢ *Anaerobic* micro-organisms (which work under anaerobic conditions when neither free oxygen nor nitrates nor sulphates are available).

All these three sets of micro-organisms are known by their common name "*decomposers*". Nature has so provided that since there are three sets of "decomposers" present in nature: (aerobic, facultative and anaerobic) organic matter degrades in any case whether oxygen is present or not. Breakdown of organic matter occurs anyway. This breakdown is essential as it releases the nutrients which are recycled to give more growth through photosynthesis. Nutrients are never lost, they are recycled, as are some other items. Later in Chapter 7 we will see more about this process as a way of treating wastes.

Depletion of Natural Resources

From Fig. 1.1 we also see that various industries, transport and public activities all use *natural resources* such as air, water, fossil fuels, metals, minerals, etc., which therefore diminish as a national resource. This subject is further discussed later below. Some of the metals and minerals being persistent go on accumulating as a waste product in the ecosystem. Like the nutrients, the persistent chemicals are never lost. But, while the nutrients help produce useful food, the persistent substances merely accumulate in the system. Some of them are even toxic and can harm health and environment when their concentration builds up sufficiently, as discussed later.

The "wastes" produced from use of natural resources are either solid or liquid or gaseous and are again either degradable or non-degradable. The degradable wastes undergo either aerobic or anaerobic breakdown as seen before, while the non-degradable fraction goes on accumulating in the ecosystem as shown. *Gaseous wastes* (CO_2, CH_4, NO_x, etc.) enter the atmosphere and some of them lead to gradual global warming and ozone depletion while as said earlier, *oxygen* (O_2) released by photosynthesis supports life.

The *degradable* wastes degrade (i.e., chemically breakdown) and release back the nutrients which along with water and photosynthesis generate new primary producers (algae, grass, etc.). These primary producers in turn become food for secondary producers which together with primary producers help to sustain tertiary ones. Humans are at the tertiary level in the food chain. The inter-dependant nature of this ecosystem keeps it going on and on.

Non-degradable wastes and wastes of long half-life accumulate in the ecosystem including the food chain and in the atmosphere with time and deteriorate the ecosystem.

Pollutants

Various pollutants enter an ecosystem mainly through air, water, minerals and other natural resources. As we have just seen, these pollutants may either be degradable or non-degradable. The non-degradable ones are also called "conservative" pollutants. Among the conservative pollutants, the biologically accumulative ones are called "persistent" pollutants.

Degradable pollutants include domestic sewage which mainly contains degradable organic matter. Industrial wastes may contain degradable organic matter plus complex organic and inorganic substances some of which may be degradable and some non-degradable. As seen earlier, degradable organics are acted upon by microbes which are referred to by their generic name "decomposers".

Non-degradable (conservative) substances are generally inert and do not degrade with time. Heavy metals such as mercury and lead are of the non-degradable type and are also biologically accumulative (persistent) tending to accumulation in the food chain. Typical other examples of persistent substances are: cadmium, arsenic and manganese. Some pesticides and radioactive isotopes can also be persistent and harmful to health.

Webs and Food Chains

Nature has found a unique way of keeping the ecosystem operational but it multiplies problems for us. This is called "bio-accumulation". It simply means that when one organism in an ecosystem swallows up another, the whole burden of pollutants contained in the victim are transferred bodily into the predator organism. Thus, when victim no. 1 is gobbled up by predator no. 1, the latter receives the pollutant load of victim no. 1, and when predator no. 2 is gobbled up by predator no. 3, the latter receives the pollutant load of both nos 1 and 2.

A classic example of this phenomenon is given by Odum (1971) with reference to the concentration build-up of DDT as shown in a tabular form below:

Item	DDT mg/L
Water	0.00005
Plankton	0.04
Minnows (fish)	0.2–0.9
Larger fish	1.3–2.0
Scavenger bird	6.0
Fish-eating duck	22.8

Build-Up in Food Chain

From the negligible concentration of DDT in water one would never suspect that it can build-up to such a high value in the poor duck which one may receive on the dining table! This is the dramatic effect of the food chain!

What is more, besides the food chain, there can be a "*web*" woven in nature. A worm at the bottom of the lake mud as well as the aquatic plants growing in a lake bottom may both be picking up a pollutant like nickel or chromium from the bottom soil. This pollutant may then pass on to algae from both these sources and from the algae to a herbivorous fish and then to a bigger carnivorous fish. Thus a web of transfers can take place. Ultimately, that fish contaminated by a web of sources may be eaten by a human being. The overall concentration factor may thus range even up to 1 : 10,000. This is somewhat like the boomerang effect where what one throws out to a lake or a sea in terms of pollution can come back to one multiplied many times over in the food obtained from it. Beware! Both vegetarians and non-vegetarians are exposed, but the latter have to fear more sources.

1.1.4 Species Diversity

When we introduce a toxic substance into the environment or raise its temperature or change its pH, we spoil its ambience. If that kills a species, we get a reduction in the number of species, an imbalance is caused somewhere and the whole ecosystem gets affected. We have to respect nature if we want to survive. Nature is unforgiving. *Many have not yet realized that in economics and ecology there is no forgiveness for those who err. In economics one loses money, in ecology your very life is at stake, slow but sure.*

The composition of any ecological community depends on the physical, chemical and biological conditions present in the environment at that time. Some organisms are more tolerant to different conditions than others. Thus, at any given time, the composition depends on the physico-chemical and biological condition obtaining at that time. The greater the variety of organisms (species) present, the more robust the system as energy can flow from a number of "channels" to sustain the system. Thus, we observe, *there is stability in diversity*. In a polluted condition, the system becomes sensitive as it has fewer "channels" of flow and an organism may succumb if food does not reach it. The whole network may fail.

If enough food is available but the environment is unfavourable, only the number of individuals of the resistant type may increase resulting in less variety of species but more number of individuals of certain species. This concept of "*species diversity*" can be used to describe the level of pollution in a natural water body. For example, an open unpolluted tropical sea may show 30–40 species per 1000 individuals examined whereas a bay may show only 5–30 species depending on the level of pollution in it, and a polluted sea may show even less than 5 species while counting 1000 individuals as only a few tolerant species may survive under such conditions. Thus, species diversity is an indirect method of measuring the pollution level in a water body.

The Carbon and Other Cycles

Various cycles operate in nature. The substances move from one state to another and back to the same state, but are not lost from the system. A few of the important cycles are:

- The nutrient cycle (seen above)
- The carbon cycle
- The energy cycle

The carbon cycle There are three sources of carbon dioxide in nature: combustion, decomposition and respiration. Combustion can be a major source of carbon release where fossil fuels (coal or oil) are burnt. Decomposition can be locally a major source of CO_2 where anaerobic decomposition of organic matter is taking place. Respiration applies to all trees/green plants, animals and human beings (about 15% of the oxygen produced by photosynthesis is consumed in respiration).

(a) Rivers

In every country, a river serves more than one city or town located on it. As the river flows past each town, the town adds its own pollution load to the load already carried by the river from earlier towns. At the same time, some diminishing of the organic waste goes on through "treatment" occurring within the river as it flows because the flowing water aerates itself and satisfies the oxygen demand of its aerobic microbes to some extent. However, the time available for "treatment" is only the time taken by the river to travel from town A to town B. This time is often not sufficient for all the organic matter to be stabilized.

In India, most rivers are heavily polluted by domestic and industrial wastes and the time for "self-purification" between two adjacent towns is often not enough. Moreover, there is a certain amount of organic load coming in all the time from the river shores (called distributed load). So the river arrives at town B often in a more polluted condition than at Town A. This water is then "treated" in a normal water purification plant and chlorinated before it is supplied to the public of town B. The water supplied hardly meets drinking water standards. This sort of half-hearted treatment can go on at many towns downstream unfortunately.

Moreover, conservative or persistent pollutants just go on building up in concentration from town to town. They do not undergo any reduction with time. In fact, the pollutants may accumulate in the aquatic plants and fish in the river, and enter the food-chain. When this sort of thing happens, one can say that the water supply is deficient in both quality and quantity. This is discussed further in Chapter 6. Fortunately, this is not the case in every river. In some cases, the incoming pollution load from a town or village is not so heavy, adequate treatment is done, and the travel time to the next town is long enough for natural self-purification to occur.

> In this sense, it is sometimes joked that Agra "drinks" the diluted sewage of Delhi, and Allahabad "drinks" the sewage of both Agra and Delhi – with all its pollutants!

(b) Lakes

Rivers usually start from mountains and end in lakes or start from lakes as they overflow and continue downstream to the sea. Hence, lakes (as well as seas) finally receive all the pollutants carried by a river. We do not worry about what the sea receives because it is already salty to begin with and no one would drink it. However, we must guard against pollution of lakes as they receive pollution from two sources:

1. The rivers that empty into it, and
2. The pollution generated by the habitation living around the lake.

Moreover, lakes have a delicate balance of their own. Lakes are relatively still compared to flowing rivers. They are known to stratify seasonally. This allows the silt and other matter carried into it by the river to settle to the bottom of the lake. But, lakes are also known to "overturn" seasonally, bringing up the settled matter to the top along with the nutrients. Thus, nutrients are everywhere in lakes. Solar energy, however, is available only in the upper portion of a lake and algal growth therefore tends to concentrate in the upper portion. Fish in the lake are of two types: surface-feeders feed on algae in the upper part of the lake and bottom-feeders feed on worms and other invertebrates at the lake's bottom.

The condition of a lake reflects what is going on around its periphery and in the catchment area of the rivers that drain into it. This is discussed further in Chapter 6 as it affects drinking water quality.

(c) Underground Waters

Underground waters are often clean-looking because they have had the benefit of natural "filtration" through the soil. Filtration is a mechanical action which removes discreet, floating impurities, but does not remove dissolved material. Thus dissolved impurities can remain dissolved in the water. This is the reason why ground waters cannot be assumed pure even though they may look very clean. Dissolved substances can be hardness causing or may contain iron and manganese or fluorides or arsenic, etc., which may need to be removed before the water is supplied for public use.

In an urban area with sewers laid in the soil to carry sewage, the upper layers of soil (2 to 4 meters deep) may always be assumed to be contaminated by sewage as all sewers are reputed to leak and contaminate the soil. If the mains supplying drinking water are also laid in this same zone and if the water supply is intermittent (as is often the case in Indian cities), they face the danger of sucking in polluted groundwater during non-supply hours.

> Ground water, even though it may look clean, is assumed microbially unsafe and thus disinfected before drinking. But, what about the presence of dissolved substances? They have also to be removed.

(d) Effect of Forestation/Deforestation on Environment Quality

It is interesting to see what effect forestation or deforestation can have on environment quality, especially water quality of a river or stream passing through such an area. Borman (1973) explains the result of a full-scale study conducted by the US Forest Service who literally cut down an existing forest over 15.6 ha area to study the water quality in a stream before and after the cutting. The study showed that deforestation led to undesirable changes in river water quality.

Since deforestation removed the tree cover, the average volume of stream flow increased by as much as 40% over the earlier value. The increased flow naturally led to increased soil erosion which resulted in a 4.3 times increase in particulate matter contained in the flow, and a 14.6 times increase in dissolved inorganic matter. Deforestation also reduced the flow of ammonia nitrogen to higher plants but increased its flow to nitrifying organisms thus increasing the concentration of nitrifying organisms in the stream. This affected the cation-exchange mechanisms of the soil and water, and the whole quality of the water was affected. It is presumed that the effect would carry on even further as the flow enters lakes and agricultural fields.

Forestation not only improves the quality of water in a stream. It also has the ability for absorbing CO_2 from the atmosphere and reducing global warming potential. Deforestation destroys this ability to absorb, and what is more, adds to the CO_2 in the atmosphere if the trees, leaves, etc., are burnt, releasing CO_2. Thus deforestation affects climate change as well as water quality downstream.

Mining operations also often lead to deforestation. Even for the simple reason of opening up more land for agriculture, a lot of deforestation occurs in our country.

We must have more forestation rather than deforestation. Western scientists tell us that each tree can absorb 1.0 tonne of CO_2 in its lifetime of about 30–40 years. Actually, trees in tropical areas with their dense growth all the year round should absorb 2 or even 3 tonne of CO_2 during their lifetime. But, this needs some field studies to show the difference between tropical and temperate region trees.

The US Forest Service has shown what a tremendous effect deforestation can have on the water environment in an area. Besides this, deforestation has a tremendous effect on the gaseous environment because it releases carbon to the atmosphere and carbon has a long half-life. What was released 150 years ago is presumably still there!

(e) Effect on Oceans and Other Water Bodies

The vast oceans and other water bodies in the world act as "sinks" for carbon emitted from different "sources" and help maintain a balance between the two. Of course, the balance is not perfect, especially in these days of increased combustion and other emission sources.

Mangroves make good sinks as they are reported to be able to absorb 50 tonne of carbon per hectare. Coastal waters have more nutrients washed down by rivers as they empty into the coastal waters. Algal growth flourishes there and acts as a sink for carbon from the atmosphere. Notice the greenish coloured coastal waters hugging the coastline next time you are up in the air. The open oceans and seas are normally saline and have alkaline pH values. They are reported to absorb about 1/3rd of the carbon emitted to the atmosphere by burning fossil fuels. As CO_2 is acidic in nature, the alkalinity of the seas is gradually getting neutralized and the pH values falling. Coral reefs formed earlier over the years are getting affected. Fisheries are also getting affected.

Sperm whales in oceans affect the situation too. Australian biologists claim that each whale excretes about 50 tonne of iron per year into the sea. This iron is good stimulus for growth of algae and photosynthesis. With an estimated number of 12,000 such whales in the Southern Ocean alone, there is a lot of scope for algal growth and consequent carbon absorption. Yet our emission is so huge, carbon is actually accumulating in the atmosphere.

(f) Global Warming and Climate Change

Finally, we find that as the world emits more and more CO_2 and as it accumulates in the atmosphere, the average temperature of the world increases. However slight this increase may be, it affects the global ecosystem. As temperature changes, the wind pattern changes. This changes the rainfall pattern and increases the likelihood of extreme weather events and global climate change.

and food chains spread the ill-effects of toxics. Again we need the use of "cleaner technologies" to reduce the use and production of toxics.

- Everything we burn or react with produces CO_2 or CH_4 or some other so-called greenhouse gas which accumulates in the atmosphere and leads to Global Warming.
- Ozone Depletion also occurs due to release of certain gaseous substances which react with natural ozone in the atmosphere.

Objectives for Control of Pollution

If we want to protect our environment, we have first to develop our OBJECTIVES for control. If our objectives are correct, our national policies and strategies will be correct. Our objectives have to take into account what we want to achieve, namely, at least the following:

1. We have to conserve all our natural resources (water, power, fuels, minerals, forests, etc.) as far as possible. There is only limited replacement of natural resources in time, by nature.
2. We should not interfere with recycling of water and nutrients in nature (through introduction of toxics, etc.).
3. Pollutants accumulate in the ecosystem as time goes, and the control standards, therefore, have to become stricter and stricter in future. Even for degradable organic wastes entering a river, the present permissible standard for final effluent BOD of 30 mg/L will have to be brought down to 20 mg/L someday and then perhaps to 10 mg/L in future as pollution loads on rivers increase with population growth and development and as river flows diminish for the same reason.
4. Development of "Cleaner technologies" will assume even greater importance in future. Our Research and Development activities will have to be guided more and more by this philosophy. We must pollute less so that we have to clean up less.
5. Global warming occurs slowly, but surely! We have to reduce the use of fossil fuels (coal and oil) in housing, transport and industry so as to reduce the discharge of CO_2, CH_4, and other greenhouse gases which are said to be causing global warming and climate change.

To achieve this objective, we have to spell out some specific *policies* which the authorities/technocrats can implement in an overall way. Further down, implementation becomes facilitated by all concerned if we spell out related *strategies* which the people can understand and follow on a day to day basis. To fulfill the stated strategies, projects and programmes have to be formulated depending on time and funds available. This, in short, is the hierarchy of planning and funding which every nation tries to follow.

The sequence is,

Objectives → Policies → Strategies → Projects and Programmes

For example, if our overall objective is to promote more wastewater treatment, we could lay down various policies to control the discharge of pollutants and protect the environment. The strategies could be many such as the use of "natural" treatment methods to conserve resources, preference for development of certain sites through availing of certain concessions, or taxation benefits, or to derive advantages from reduce-reuse-recycle, etc.

If we are not satisfied with either the pace or the specific strategies we are following, we must first review them. Defective or inadequate policies can be the root cause of many a problem. The pace at which things are happening generally depends on the funding available and the work-culture of the people. It is no use merely increasing the funding if the direction of progress is itself wrong. No use hurrying in the wrong direction! First, make sure the policies and strategies are in the right direction!

This rule applies to almost everything that one can think of. As we go along dealing with all the other topics listed in this book, we have to always see if the on-going policies, strategies and programmes all have been spelt out logically and sensibly. If the first three items are spelt out logically, the Programmes and Projects that follow will also reflect the same trends.

(A discussion may be held to see if the objectives will be met if the policies laid down are followed, or still more need to be added or some changed. Similarly, for satisfying these policies, are the strategies identified adequate or different ones need to be spelt out? Finally, are the programmes and projects planned or in hand based on the strategies identified or is there a disconnect between the two. Different viewpoints will be expressed and will need to be reconciled taking help of Figs 1.1 and 1.2. See how the situation can best be remedied without loss of previous efforts and resources as far as possible?)

1.3 ENVIRONMENTAL QUALITY ASSESSMENT—USE OF CRITERIA AND STANDARDS

> Evaluating the environment is serious and complicated business. It involves water, air, wastewater, industrial waste, solid wastes, hazardous wastes, and many more items affecting daily life.
>
> If you cannot evaluate something, you cannot value it, nor can you improve upon it.

Environment can affect the livelihood of many people and generate much resistance to its enforcement if perceived to be unnecessary. It can often become a political issue if not handled expertly. Evaluation has always to be firstly to protect human health. Health surely comes first. But, as the World Health Organization (WHO) likes to remind us our objective should be broader, not only to protect and promote health, but also health and well-being. Thus, the definition of health must include peoples' well-being too.

Public health is directly affected by several agents among whom the following are most important:

- Microbes, flies, mosquitoes and other organisms which can give diseases (e.g., malaria, filaria, dengue, typhoid, cholera, dysentery, diarrhoea, etc.)
- Persistent substances (organic and inorganic) which can enter the food chain and affect health (e.g., mercury, cadmium, pesticides, etc.)
- Physical items, like noise, heat, radiation, etc. (affecting health and well-being)

1.3.1 Personal and Community Hygiene

At the outset, a word should be mentioned about the generally poor level of personal hygiene observed in India. Probably out of ignorance, personal hygiene is sadly neglected. Hands are often not washed after performing various activities which soil the hands, including use of toilet. Any water that "looks" clean is considered fit to drink, especially if it is "flowing" water. People who live in slums (and half of Mumbai does!) consider personal hygiene a luxury which they can ill-afford in the absence of a copious supply of water.

A study undertaken among poor urban people showed that washing hands with just plain water reduced the infection rate by nearly a third of the cases, while washing with soap and water reduced them by more than a half. Such are the benefits of washing, with or without soap. The bugs will take care of themselves if we take care of them first!

Besides personal hygiene, lack of community hygiene (what we often call lack of "civic sense") is another problem. Indians have a well-known habit of cleaning up their houses, shops, establishments regularly and well, but dumping the waste on the adjoining roadsides, hopefully, for others to clean up. Often, the wastes end up blocking open drains. Plastics fly about. Garbage is chewed upon by animals and mosquito nuisance is rampant – leading to a variety of health problems. This is clearly an environmental health problem with roots in the peoples' culture.

Observation: Not only is personal hygiene poor but community hygiene is even poorer.

Organizations like *Sulabh Sauchalaya* in India are well known for their work. NGOs generally recognize that solving the sanitation problem is more than setting up latrines. It has to be backed up by educational and training programmes that are effective at the local level.

Why should All Citizens Worry about Public Health?

As listed below, a large number of microbes and pollutants can be present in water at any given time. Senior citizens and children are most vulnerable to infection because of their age. Slum dwellers are most exposed to infection because of their location.

There is, however, often a time lag between cause of illness and on-start of the illness. Some infections may take up to a week or two after the contaminated water is drunk or after a mosquito bite occurs, before actual onset of illness starts. The effect may occur even much later in some cases after years (e.g., cancer) – so the connection between the two may be missed. This is a worry.

Some Pollutants and Case Histories

Several hazardous pollutants and microbes exist in our environment such as the following revealed by a few famous case histories:

data is important to set the Acceptable Daily Intake (ADI) levels of mercury.

In an average brain size of 1.6 kg, Japanese studies show that the Mercury contained at time of death = 1600 g × 5 ug/g = 8 mg. If this constitutes 10% of Total Body Burden (TBB), the latter would be 80 mg.

2. As the daily elimination rate is 1% of Total Body Burden (TBB), the mass eliminated daily = 0.8 mg. At steady state, the intake has to be limited to the same amount going out of the system (0.8 mg).

   However, other epidemiological evidence from Japan shows that for *susceptible* individuals the limit is only 0.3 mg instead of 0.8 mg. Thus, this data should be used rather than the previous one. Applying a safety factor of 10 to this amount, the Acceptable Daily Intake (ADI) = 0.03 mg/day. This is called "criteria". It is applicable universally to all people in all countries. It has a safety factor of 10 already built in.

Standards are Different from Criteria

Standards are different from criteria. Standards are applicable to each specific country. Standards are a political decision based on people's eating habits and the risks and costs involved with the type of food available in the country. Standards can be different for different countries. *Criteria* are absolute values which apply to all countries. Standards are also applied to the more easily measurable day to day items in the market of that country for control purposes.

For ease of daily control on fish sold in the market, the above criteria value of ADI, namely 0.03 mg/day is thus converted to mg of mercury permissible per kg of fish sold in the market. This requires *sociological studies* at each country level to determine mercury levels found in local fish and actual fish eating habits of the people. Standards are thus developed country-wise as shown below.

For example, the standard set in Sweden based on their dietary habits, (which are more fish-eating) is 0.2 mg/kg of fish flesh so that the mercury intake does not exceed the criteria value for the average local population. Another country may relax this limit to 0.5 mg/kg fish flesh because a stricter standard may cause a politically unaffordable risk of unemployment in a certain section of the fishing community, knowing full well that it would cause a risk to health of heavy fish eaters! Hence,

Standards are a political decision of each country. In fact, such standards are called *derived* standards.

Incidentally, one should know how the standards for different pollutants have been derived for one's own country. One should at least know how to secure information on this subject from the pollution control authorities.

1.3.3 International Organizations

International Organizations most referred to while setting criteria and standards at a country level are:

- ➢ World Health Organization (WHO)
- ➢ Food and Agriculture Organization (FAO)
- ➢ International Atomic Energy Association (IAEA)
- ➢ International Maritime Association (IMA), etc.

> Remember pollutants accumulate in the ecosystem. Thus, standards in all countries need to be updated, and are generally made stricter with time.
>
> Review of standards is, therefore, a periodic activity for any country, not a one-time affair.

1.4 NATURAL RESOURCES AND NEED FOR CONSERVATION

Natural resources include at least the following:

- ➢ Water
- ➢ Land
- ➢ Forests
- ➢ Fossil fuels (coal, oil)
- ➢ Minerals, and
- ➢ Chemical deposits, etc.

Some of our natural resources are *renewable* (such as water) and some are *non-renewable,* or relatively slowly renewable over a period of time (such as the fossil fuels like oil and coal, chemical deposits, minerals, etc.). All natural resources must, therefore, be used carefully (conserved) as best as possible so as to keep in balance with their natural (slow) renewal rate.

Conserving any natural resource is firstly a matter of knowing what activities are involved in its production and use, going backwards from the product to the source. For any product one has to trace, so to speak, from the "cradle to the grave", the activities involved and the resources used and how renewable or non-renewable they are, and how best they can be conserved.

Based on this knowledge, appropriate policies can be laid down by government including its pricing policy. Proper pricing is important otherwise decisions would be based by users on market availability and not much regard given to other aspects.

1.4.1 Water

Why is it essential to conserve water? Why has conservation not been very successful in the past in India?

About 97% of the total water available on earth is in the oceans. Of the remaining 3%, only 0.003% is available to the world as fresh water in rivers, lakes, etc., the rest being frozen as snow. Of the fresh water available to the world, only a small fraction is available to India and has to be used carefully. In India, water is a *state* subject. Each state is free to handle its own water resources. In allocating resources, each state generally gives first priority to drinking water, second to water for agriculture and then to water for industry.

The water resources of India are reported to be used for various purposes, roughly, as follows:

- Agriculture 80%
- Municipal and other uses 10%
- Industry and commerce10%
- No State seems to give any priority to water conservation.

Why is Water Conservation Essential?

The amount of fresh water available per person in India is rapidly decreasing. Why?

Rainfall as a source of water is limited and per person availability of water in rivers is decreasing for the following reasons:

- Population in the catchment area is increasing rapidly.
- Agricultural and industrial usage of water is also increasing rapidly.
- Lifestyle changes are taking place.
- The water we pollute becomes unavailable for beneficial uses.

Hence, in all countries, water conservation becomes essential as time goes. However, conservation and reuse are considered, only as a last resort.

Just as necessity is the mother of invention, so also water conservation is done by people only when fresh water becomes either:

- ➢ Unavailable, (scarce) or
- ➢ Unaffordable, (expensive) or
- ➢ Undrinkable (polluted)

In some Indian rivers today, the per capita availability of fresh water is only 1/2 to 1/3 of what it was just 30 years ago! This is a true fact.

Mumbai city was earlier supplying drinking water @ 150 L per person per day. Today, the city can hardly supply 90 L per person per day. It is already talking of reducing it still further. *Rainwater harvesting* in cities like Mumbai is actually not a good idea from the health viewpoint (see Chapter 3) but it is recommended by the local municipal corporation to overcome shortage of supply.

Thus, Water Conservation has to Become a Part of Our Lives

Irrigation people use 80% of the country's water resources. Yet, they have little to show for conservation. They have two opportunities to conserve water: (1) drip irrigation, and (2) lining of conveyance canals. Both are useful but expensive and omitted as long as water is easily available!

Drip irrigation has transformed Israel who first developed it in 1969. Their agricultural output has doubled since its introduction. Drip irrigation applies water directly at the root zone of each plant through a specially designed water "emitter". Thereby, under-drainage loss is avoided, unlike basin or contour irrigation where only 50% or less of the applied water is actually used by the plant. However, to prevent emitters from getting choked, the irrigation water may need some pre-treatment. Drip irrigation systems are manufactured and are increasingly used in India too.

In case of unlined canals, much water may be "lost" underground in transporting it so that not only water is lost but the adjacent soil may sometimes be ruined by "salination". The percolated water evaporates leaving its salt behind which gradually turns the soil saline.

Strangely, sometimes political compulsions act against water conservation and economic reasoning. For example, some authorities in India offer *free* electric power to farmers for pumping of ground water. This tends to lead to over-use of water, and acts contrary to the principles of conservation.

Reuse of Wastewater

Reuse of wastewater is another effective form of water conservation. However, treatment before reuse water costs money. Hence, reuse is generally avoided as long as fresh water is available unless it is a mandatory requirement. Relative costs and other important factors discussed in Chapter 8 determine the situation.

Low Price of Water in India

In a relatively poor country like India, for obvious reasons, price of water is kept unduly low (we charge ₹ 3/- for what costs us around ₹ 15/- to supply) and give farmers free electricity, with the result conservation becomes unimportant. Groundwater is often over-abstracted and water quality is not assured.

In a city like Mumbai, in the year 2010, 20% of municipal samples *daily* fail bacteriological tests! And you get a free cocktail of pesticides too! Both quality and quantity are gone and we are at the mercy of others for more funds.

Other Consequences of Low Pricing

Resources earned from sale of water are not adequate for timely renewals and additions to infrastructure. The country has to depend on external funding for its basic infrastructure.

Cross-subsidy with industrial water is commonly done but it makes the latter expensive.

Some wastewater is reused in industry and agriculture. Yet, communities as a whole in India do not regard wastewater as a reusable "resource" for augmenting public supplies. Wastewater is just wasted! (see later for Reuse possibilities).

Suggested Linking of Rivers in India

A novel way has been suggested in India of ensuring that rivers in the south of India be fed with surplus waters from the relatively water-rich rivers of North India, by linking the northern rivers with the southern rivers. This idea has been very actively favoured by many authorities dealing with water resources in India. However, one wonders if this will truly end up as a panacea for water shortages in Southern India, or cause some unexpected problems?

Perhaps from a quantity viewpoint, this may prove to be a great (though expensive) idea. However, from a quality and ecosystem viewpoint this may prove a disaster which only time will tell. The waters of the North are generally snow-fed and therefore cold. Their introduction into the warmer waters of the Southern rivers will perhaps cause a kind of thermal pollution in reverse and affect the microorganisms unfavourably. The whole ecosystem will no doubt eventually adjust to the new temperature and food regime with several species (perhaps some of commercial value) may be lost in the transition. This is what we do not know. All we know is that there will be an effect and the effect will cause a ripple of other effects to follow. The case of the US experiment of deforestation and how it affected water quality cited earlier (Section 1.1) may also be recalled. A similar situation will arise in India. Who knows whether linking of rivers in India will after all prove to be a good idea or an ecological disaster waiting to happen! When the disaster happens, it will be too late to retrieve our steps. Conservation and reuse seem a better idea to begin with.

So If You Want Conservation, Give the Right Signals

- Water conservation must be strictly enforced wherever rainwater harvesting is being done because both give reduction in primary demand for water.
- Where rainwater harvesting is practiced the water may be stored by direct storage in a tank or, if soil permits, as groundwater recharge. Of course, this should not compromise consumers' health as discussed in a later Section.
- Do not give free electric power for pumping as it gives wrong signals and also tends to encourage overuse and/or wastage of ground water.
- Where some subsidy becomes essential, keep it to a minimum and give it stepwise to avoid misuse. Be very focused. Avoid cross-subsidy as far as possible.
- Ideally, price the water such that the price charged is more or less equal to the cost price of supplying water.
- The water authority should be able to generate as far as possible their own resources to undertake timely expansion of water projects in future.
- Encourage reuse of water, especially for industries and for certain municipal purposes.
- Local bodies have to promote water conservation constantly in the public mind. It has to become a routine exercise for them.

- Spread conservation awareness among public, especially of groundwater. Groundwater usage is often left to the user's judgement, and he generally tends to over-use.
- Prevent leaking taps, pipes, fixtures and fittings (both public and private).
- Insert orifices to reduce pressures and flows wherever necessary.
- Install dual-flush systems to reduce water consumption or use pour-flush toilets.
- Control pollution of all water resources so as to ensure maximum availability of fresh water for legitimate consumption.

> In USA, they are clever, they say: "WATER IS FREE, WE ONLY CHARGE FOR PRESSURE AND PURITY !"

1.4.2 Land

Land is also a natural resource and its use must be made with care. It is primarily used for agriculture and construction of housing, transport and infrastructure. Industry usually comes later and buys the local land. Land is also "renewable" in a sense, as its use or reuse may change with the owner and the structures built on it may be renewed. Fresh use of land is generally made when land ownership changes or the old structures on it have to be rebuilt/renewed.

Natural methods of wastewater treatment such as oxidation ponds, constructed wetlands, etc., always need more land per person than mechanized methods of treatment such as the activated sludge or trickling filters, but they have the advantage that no machinery nor electric power is needed to operate them. Thus a conscious decision has to be made on their use depending on land availability and price.

Large areas are often required for installing wind and solar energy systems as also nuclear energy systems (Chapter 2).

Government always arms itself with the right of acquisition at a special price. This right is normally to be used for important public infrastructure purposes. Misuse is sometimes attempted for political purposes or to favour a project or an industry. The recent difficulties experienced in Singur and Nandigram in West Bengal show that politics is often involved. This is nowadays vigorously resisted by local politicians who intervene and cause trouble. Hence, industry often in its own interest leaves out the government and finds it better to negotiate directly with the land owners and settle on mutually acceptable terms.

Experiences elsewhere indicate that it may be advisable to make the payment in a dual manner: cash and kind, so that even if the cash component is frittered away by the ignorant farmer, the other part will see him through. New legislation is being envisaged at present in India to protect both the farmer and the industrialist.

In recent years, efforts are being made to get over the land problem in various ways. For example, wind energy can be harvested on the open sea instead of land, solar energy can be harvested on satellites circling the earth, nuclear energy is being continually made much safer to use, and to top it all, measures are being taken to "create" new land (like what the Dutch and Dubai authorities are doing) in the form of new "islands" created by digging up the sand from the sea bed and depositing the same at high-value locations elsewhere. However, this kind of activity is expensive and can only be justified in a few cases. It is more important to conserve and make careful use of the land which nature has already given us.

1.4.3 Forests

Forests constitute another natural resource which needs protection. Deforestation is often attractive to unscrupulous traders on the look out for wood and other forest products of commercial value. Deforestation may also be done with the intention of creating open land for sale for agricultural or other activities. Their control requires better counseling and sometimes better policing too. The reader is referred to earlier discussion in Section 1.1 on how deforestation can change water quality in local streams and also affect global warming.

Deforestation is also done unwittingly by womenfolk in search of wood and twigs for their domestic cooking purposes. This is best controlled by providing them with alternative fuels such as Gobar Gas for cooking. Such plants may be based on human or animal wastes locally available.

In recent times, one more reason has become apparent for preserving forests and developing new ones where feasible as explained in Section 1.1. The reason is to counter global warming. The carbon present in the atmosphere is diminished by absorption by trees during photosynthesis. So, it is extremely beneficial to have as many forests as possible in a country.

1.4.4 Fossil Fuels

The world mostly uses fossil fuels like coal and oil for combustion purposes at present. Fossil fuels are reaching their exhaustion limit and predictions are aplenty that within the next 2 or 3 decades the world will run out of its stock of fossil fuels. This is perhaps all to the good because their combustion gives carbon containing gases (greenhouse gases) which are said to be causing global warming all over the world and changing its climate as we know it. Sea levels are also rising owing to snow melting. Moreover, India has to import 80% of its oil requirement – which it can ill-afford.

This subject is important for all of us, and alternative sources of energy are discussed in the next Chapter on "Energy Production".

1.4.5 Conservation of Naturally Occurring Minerals, and Chemicals

Similar to the above cases, we need to see what minerals and other natural substances we have in short supply. Then, check what our policies are in its regard. With recent extensive demand for certain metals used for manufacturing chips and other parts for mobile phones, computers, etc., policies are urgently needed to conserve such metals and minerals.

For example, copper has always been in short supply in India. That is why unlike European countries, we have generally not used copper pipes for plumbing. However, now with imports easily available, copper is being increasingly used where market prices are favourable. Globalization encourages its use, but, for strategic purposes (during conflicts, when copper may not be easily importable) policy constraints may have to be imposed.

Global warming considerations should also put a constraint on anything that has to be transported over long distances. So we may sometimes have several considerations to take into account in formulating our policies.

1.5 THE PHILOSOPHY OF WASTE DISPOSAL—SOME INDIAN STANDARDS AND ENVIRONMENTAL LAWS

One may ask: "What is the basic philosophy that guides public policies framed in India in regard to water supply and waste disposal? The answer would perhaps be: "These are based on an ecosystem approach" as illustrated in Fig. 1.1 which is a very important diagram as it illustrates the ecosystem approach and provides us with a scientific basis to formulate our policies and programs wisely. The Indian standards and environmental laws are based on this philosophy!

The basic objectives behind this approach are:

- Environmental health protection, and
- Resources conservation

In dealing with both wastewater and solid wastes, the philosophy is based on a simple thought: "Waste is not to be wasted". The interesting thing we see from the ecosystem diagram is that the "waste" of one becomes the "food" of the other. Nature seems to have created organisms because "food" was there. We have to do likewise. Waste is not to be wasted. It must be treated, used and reused as nature always does.

Actually, much work has already been done in India and many strategies are already being followed for enhancing environmental protection and conservation such as:

- Regulatory measures
- Legal requirements
- Fiscal incentives
- Technical and scientific developments

Government has established expert bodies in each of these matters, and based on experience, changes are made as required.

1.5.1 Approved Procedures

When a new industry has to be established or a new urban development is to occur, the project proponent is generally required to undertake an Environmental Impact Assessment (EIA) of the project. They are generally expected to study the area and the local availability of water, land, fuel, labour, etc., and make recommendations about the suitability of the site. The site and the industry must match. This is discussed further in Chapter 5.

The industry is expected to use "clean technologies" in the manufacture of whatever product it proposes to make, and preferably a Life Cycle Analysis (LCA) of the product is expected to show the following:

1. Use of minimum resources will be made in developing its infrastructure the form of power, water, land, etc.
2. Use of the least polluting technology will be made in manufacture.
3. Production of the least amount of waste will occur in its manufacture, use and final disposal.

ecosystem and the consequences of not following them. The ecosystem diagram is not always followed especially when money is involved. They must also understand concepts like "*sustainability*" and "*resource conservation*".

The people (not just the technocrats) must see the reasons for action and want to take action themselves. We generally tend to preach at government employees, technocrats and NGOs in conferences and meetings. But, we have to be able to teach the common man. This takes time. Often, lot of time and effort! We are hopefully beginning with young engineers who read this book.

Community empowerment Capacity building has to be followed up with community empowerment. Community empowerment *facilitates* the people to take action. How?

1. It gives them the "right to information" and the right information so that they may take the right action.
2. It involves them in decision-making so that they get the feeling of ownership.
3. It ensures transparency of action.
4. It develops proper authorities, expertise and facilities to certify goods and services.
5. It encourages NGO activity and activism in general and Public Interest Litigation (PIL) also where necessary.

When capacity is built and the community is more fully empowered, things will happen quickly. Let us examine our existing policies, if any, for achieving capacity building and community empowerment. Have we spelt out specific strategies for achieving them? Do we have any on-going programs in this regard? What should we do to get the right things to happen speedily?

> Although we have already done a lot, we have yet to do more in capacity building and community empowerment to get the "iceberg" of inaction to melt and the general environment to improve. Hopefully, we are on our way.

1.6 THE ROLE OF STATISTICS

Statistics is helpful to all engineers and scientists. Civil engineers find it useful in dealing with massive data on rainfall, run-off, floods, etc.

Mechanical, electrical, chemical and civil engineers find it useful in handling various data concerning their respective fields. All engineers have to compare two sets of data, two series of data, one control and the other experimental, to see if there is a statistical difference between the two or not. At the outset, this short note on the role of statistics is included here as most young minds (and even some old ones) are unfortunately taken in by whatever is put out as "statistically valid information".

As someone said a long time ago: "there are lies, damned lies, and there are statistics". Statistics are wonderful as they can compress a lot of data into a few meaningful numbers which can tell the story better. Statistics are also useful in analyzing the degree of variance. This makes statistics a useful tool in designing new experiments and checking on the validity of other peoples' lab work. However, statistics are also sometimes quoted to promote a product or to discourage a rival's product or to baffle a person and perhaps to fool him into believing something.

Firstly, therefore, when confronted with statistical evidence, it is perhaps wise to ask if the finding is a plausible finding. There should be some independent supporting evidence for the finding, not something just out of the blue. Secondly, where is it published? Is it published in a respected and peer-reviewed journal that one can generally depend on? If still in doubt, check the source of study, who did it, and for how long was the study done, and the sample size tested – and above all who paid for it? Was it a very short duration study with a small and unrepresentative size of sample? Or, are we comparing apples with oranges? These are all commonsense questions which one can ask oneself. Statistics have a way of mesmerizing the casual reader.

Statistics says that anything has a certain chance of happening. There are two broad classes of errors: (1) Determinate or systematic errors like zero-error on a weighing machine. It affects all results equally. It can be reduced by calibration and by following standard procedures, and (2) Random errors, whose magnitude and direction are not predictable. Such errors can be treated by statistical methods.

A few simple definitions are necessary to give at this stage:

Samples of air, water, wastewater, etc., are normally drawn for testing. A different set of statistical rules apply to replicate samples and routinely drawn samples.

Replicate samples: Different tests done on the same sample.

Routine samples: Tests done on different samples (like daily water/air samples).

Some of these definitions are also illustrated graphically in Fig. 1.3.

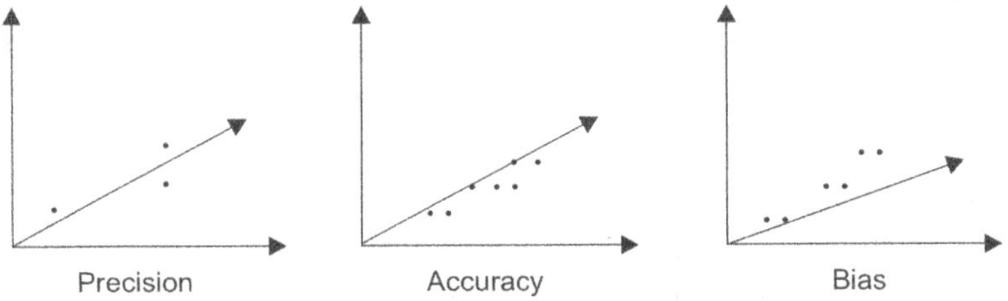

Fig. 1.3 *Some definition sketches in statistics*

Precision signifies that reproducible results can be obtained from test to test.

Accuracy signifies how near the results are to the true value.

Bias refers to magnitude of deviation of measured value from true value.

NOTE: A result may be of high precision but yet totally inaccurate because there is a bias in the measurement instrument! Of what use is such a result?

Recommended Additional Study of Statistics

As all engineers and allied personnel have to continually use statistical concepts in planning, designing and evaluating the results of various studies in air and water pollution control in the course of their work, the subject of Statistics is highly recommended for additional reading so as to gain a working knowledge of the same. One should at least be able to solve some typical problems facing engineers, such as:

1. Number of samples needed to be taken. The smaller the spread, the lesser the number of samples needed.
2. Comparing two sets of data to see if there is a statistically valid difference between the two sets.
3. Evaluation of different sites from which samples may be drawn to select the site which gives the least variation.
4. Dilution and other problems.

Observation: This was a very diverse chapter. It provided the background to policy-making in the environmental field to protect our health and well-being and our fragile ecosystem. It told us that water, land and many other items are our natural resources, but are scarce indeed, and have to be used carefully.

It showed us how much data was needed for evaluating the health effects of pollutants in food and water, and therefore the standards and codes of practice formulated for us have to be observed by us at all times.

REFERENCES

Arceivala, S J, (2010), *Global Warming and Climate Change*, Indian Environment Association.

Arceivala, S, Asolekar, S, (2007), *Wastewater Treatment for Pollution Control and Reuse,* 3rd Edition, Tata McGraw Hill.

Borman F H, (1973), US Forest Service, "An Ecosystem Approach to Problem Solving", NATO Conference.

Fair, Geyer and Okun, *Water and Wastewater Engineering*, Vol. 2, John Wiley & Sons.

Metcalf & Eddy Inc, (1990), *Wastewater Engineering*, McGraw Hill.

Odum, E, (1971), *Fundamentals of Ecology*, 3rd Edition, W B Saunders.

EXERCISE QUESTIONS

1. Show how environmental management is ultimately based on ecosystem management. Do you have any aspect of the ecosystem unclear in your mind?
2. Comment on the seemingly magic situation denoted by the following statement that "Nothing comes from nowhere, everything goes somewhere!"
3. Why do you think the environment scenario in the country is still uneven? How can we increase the speed?
4. List any five principles of environment management which you think are important. Explain briefly their importance with examples.

5. Write a brief note on Indian laws for environmental protection. Refer to the internet for desired information.
6. Comment: "If you want conservation, give the right signals" and charge the right price!
7. Distinguish between: Objectives, Policies, Strategies, Projects, Programs.
8. Which policies in your opinion are not yet covered in your State? What is the procedure for establishing new policies in your State?

PROJECT/ASSIGNMENT TOPICS

1. As an expert to whom the recent oil-spillage in Mumbai harbor is referred, describe its various likely effects on the local ecosystem components and suggest ways of combatting the same. Refer to the internet if necessary.
2. We list below a few "soft" and "hard" options for action. Soft options are generally easy to implement, but not much action may be achieved. Hard options are more difficult to implement but some useful results may be obtained. Discuss each of these options stating which ones you would recommend in each category, and why? Be brief.

   Also discuss the following options one by one for their feasibility and cost (high, low, medium) to the nation, and in some cases the unwanted results they may bring.

   Soft Options

   - Monitor industrial emissions and discharges
   - Appoint more monitoring staff
   - Levy more stiff fines
   - Legislate water conservation rules
   - Hold seminars and conferences

   Hard Options

   - Change Government policies affecting environment
   - Set up system of permits for groundwater abstraction
   - Prepare new legislation for land acquisition for public purposes
   - Prepare a list of negative subjects for award of import licenses
   - Control use of natural resources
   - Develop and implement a program to train the trainers

2 CHAPTER

Environmental Aspects of Energy Production

2.1 Emissions in Traditional Power Generation with Fossil Fuels

2.2 Emissions from Industry, Commerce and Services

2.3 Emissions from Livestock

2.4 Emissions from the Transport Sector

2.5 Important Alternative Sources of Energy

2.6 Alternative Sources Already in Use

2.7 Alternative Sources of Energy Ready to Use

2.8 Alternative Sources of Energy Needing Development

> *The traditional sources of energy have been fossil fuels. The future belongs to those who can use renewable sources of energy economically.*

2.1 EMISSIONS IN TRADITIONAL POWER GENERATION WITH FOSSIL FUELS

Energy is needed for everything we do. Energy has to be produced to support all our activities. All people need to use energy. Production of energy using traditional methods affects the environment, our health and well-being as also our resources. Hence, it is essential for us to learn about these effects and do something to mitigate them.

Traditionally, we have been using *coal or oil* to produce electricity in thermal power plants to operate our equipment and appliances. For transport we use diesel oil or petrol (refined from oil) to operate our vehicles. Thus, our society has grown on these fuels which are called *"fossil fuels"* because they were originally formed underground after millions of years exposure to heat and pressure applied on trees/wood, algae, etc., destroyed and submerged centuries ago during natural calamities (see Fig. 1.2).

What took millions of years to form, we burn in a few minutes and produce carbon dioxide (CO_2) as an end-product. The world releases more than 30 billion tonne of CO_2 per year (more than 60,000 tonne per minute). The carbon continues to exist in the atmosphere for many years. It has a long half-life of 100 to 150 years. CO_2 from the atmosphere is again used in photosynthesis to generate new plants and trees which may someday become coal for us to burn. The carbon cycle is thus completed after many millions of years but meanwhile at the fast rate at which we are burning our fossil fuels, we are more than likely to run out of coal and oil within the next few years.

Moreover, as we said earlier, the CO_2 we produce has a long half-life. Hence, it accumulates in the atmosphere. What was released 100 or 150 years ago is still there. Presently, CO_2 has built-up to an average of about 380 ppm in the atmosphere and is beginning to show its effect in the form of "global warming" which is triggering climate change. Scientists are afraid that when the world's average warming rises further by more than 2°C (probably by the year 2050) the climate change may become irreversible and the whole ecosystem will be in a state of dysfunction.

2.1.1 New Phenomenon to be Faced

For years we did not realize that the carbon dioxide, methane and other gases, which we collectively call greenhouse gases (GHG), released by us from combustion and other activities were accumulating in the atmosphere resulting in global warming and leading to a slow change in our climate. The actions we have to take are fortunately the same as those needed to protect our environment. So, in a way, we know what is to be done, and as in the case of the environment we realize that all people will have to do something.

> The danger is that what is everybody's business can become nobody's business after all !

Global warming and climate change will have a serious effect on health and sea level because of ice melting at the North and South poles. India has a long coastline and some of its largest cities and busiest harbours are on the coast. Already Mumbai has had to set-up its first pumping station in Haji Ali area to pump its storm water to the sea as it will not drain by gravity any more at high tide. Also, as the sea level rises, it "eats" inwards into the land and affects agriculture and land-use at the coastal edges. It also turns ground waters brackish and undrinkable and encourages mosquito growth.

Hence, we must review our use of fossil fuels to produce the energy we need at present and will need in the near future. We must now emphasize the need to explore alternative sources of energy. These alternative sources of energy will have to be affordable, sustainable and not generating any carbon.

2.1.2 Some Emission Factors in Traditional Power Generation

Emission factors are important to know as we cannot make any emission inventories (estimates) without them. They are generally derived from stoichiometric equations expressing combustion of the substance in question. Their value naturally depends greatly on the original composition of the substance. Coal, oil, waste agricultural products may all vary from specimen to specimen. Hence, the corresponding emission factor may vary within a range of values.

- ➢ Burning coal to generate steam and thence to produce electricity through steam turbines, gives 2.00–2.46 kg of CO_2 per kg of coal burnt (depending on type of coal).

- ➢ Burning diesel oil to generate steam and thence electricity gives 2.63–2.97 kg of CO_2 per kg of diesel oil burnt.
- ➢ Using 1 kWh of electricity produces 0.37–0.48 kg of CO_2 depending upon the type of fuel-oil used to produce electricity.

> As an approximate thumb rule (slightly on the higher side) one could say that using 1 kWh of electricity leads to an approximate emission of 0.4 kg of CO_2.

For a few other fuels:

- ➢ Burning 1 kg of LPG gas gives 0.15–0.19 kg of CO_2
- ➢ Burning 1 kg of kerosene gives 2.518 kg of CO_2
- ➢ Burning 1 kg of wood gives only 0.17 kg of CO_2 but it gives some carcinogens (so beware!)

Fuels derived from oil are theoretically less carbon-rich and contain two hydrogen atoms for every one of carbon. Because hydrogen produces more heat than carbon when burnt, burning oil releases less CO_2 per unit used than coal, and in doing so produces only water. Methane has four hydrogen atoms which makes it the fuel of choice for us.

The efficiency of burning a fuel is also a factor in production of CO_2. Burning anthracite to generate electric power results in 67% more CO_2 emission than does burning methane, while brown coal which has more moisture and impurities produces 130% more. Thus, the kind of fuel used by a country makes a lot of difference to its carbon emissions. Peat and brown coal emit the most carbon, bituminous less, and anthracite lesser still:

Peat → Brown coal → Bituminous coal → Anthracite → Diamond?

Power plant emission rates depend on type of fuel used. Their emission rates in terms of tonne of CO_2 per MWh of power are: Lignite-1.3, Coal-1.05, Diesel-0.65, Naptha-0.64, Gas-0.42.

Example 1: A family's electricity usage averages only 300 units (kWh) per month (3,600 units per year). Estimate the CO_2 produced per year by the family. Assume the power plant serving the family is gas based.

$$CO_2 \text{ produced} = 300 \text{ kWh/month} \times 12 \text{ m/yr} \times 0.40 \text{ kg } CO_2 / \text{unit}$$
$$= 1{,}440 \text{ kg/yr}$$

If the power plant is coal-based, the CO_2 emission would be higher (3780 kg/yr)

If electric transmission losses between the power plant and point of usage are say 70%, the CO_2 produced be 70% higher in each case.

2.2 EMISSIONS FROM INDUSTRY, COMMERCE AND SERVICES

The emissions from these three sectors vary widely from city to city and need to be determined carefully for each city since considerable changes have been occurring in Indian cities. Just a few years ago, Mumbai was a textile centre. Today, it is not. It is now more of a services centre. All industries, commerce and services consume electricity and other fuels. Some consume chemicals and use processes which release carbon. The emissions can be estimated approximately from the amount of different fuels burnt in local power houses as well as in private industries for the purpose. Considerable emissions also occur from commerce and services in the form of transport.

UK data gives the CO_2 emissions from industry, commerce and services as varying from: 0.6–2.0–3.0 tonne / person / year. Such information does not exist for India. However, for rough estimation purposes in Indian cities one may use the lowest value (0.6) from the above range.

For India, out of the total CO_2-equivalent emission from industries of 412 million tonne per year in 1997, nearly 32% were from mineral industries, 28% from metal industries and about 8% were from chemical industries and the balance from other industries such as pulp and paper, food and beverage, textiles and leather, mining and quarrying. Today, the percentage distribution would most likely be different.

Specific emission factors (EF) applicable to a few industrial processes are given as follows:

- *Aluminium:* EF = 1.65 tonne per tonne Al produced.
- *Cement*: EF = 0.5370 tonne of CO_2 per tonne of clinker used (estimated '94).
- *Iron and steel*: EF = 1.72 tonne CO_2 per tonne production (OHV).
- *Nitrous oxide (N_2O):* emission from Nitric Acid production depends on operating pressures. The mean value of N_2O produced in a medium pressure plant was 10.13 kg per tonne of nitric acid, while in a high pressure plant it was only 2.84 kg per tonne. It could be brought down to as low as 0.4 kg N_2O per tonne of nitric acid by use of a non-selective catalytic reducer.

- *Methanol:* EF = 0.67 tonne CO_2 per tonne methanol produced. (Also, 2.3 kg methane per tonne methanol produced).
- *Ethylene:* EF = 1.73 tonne CO_2 per tonne of ethylene produced. (Also, 3.0 kg methane per tonne methane produced).

More values are available from Ministry of Environment and Forests, New Delhi.

2.3 EMISSIONS FROM LIVESTOCK

In India, 70% of the rural population owns livestock. Livestock includes cattle, buffaloes, horses, sheep, goats, donkeys, camels, poultry and many more. It also includes their dung.

Enteric emissions from cattle (actually consisting of methane generated in their stomachs) vary with type of animal and their feed. The IPCC default values for dairy cattle are 46 kg CH_4/animal/year while for non-dairy cattle they are half as much. For buffaloes they are around 55 kg/animal/year. The methane has, of course, to be converted to equivalent CO_2. 1 tonne of methane is equivalent to 25 tonne of CO_2 in its global warming power.

Cow belching (enteric emissions) can also be reported in terms of volume @ 600L of methane gas per day per cow. (In terms of equivalent CO_2 it is an enormous amount of about 15,000 L of gas expressed as CO_2-equivalent per cow). It is clear that a country like India which venerates the cow, keeps giving off a large amount of equivalent CO_2 till the animal dies of old age. In the western countries the animal is slaughtered for its beef after it stops giving milk.

Observation: As we said earlier, the world produces an average of more than 30 billion tonne of CO_2 per year – of which India produces about 2 billion tonne per year. Both India and the world will produce more as the years go by. Hence, we must examine carefully what options we have for using alternative sources of energy producing low carbon so that our growth potential may not be required to slow down.

2.4 EMISSIONS FROM THE TRANSPORT SECTOR

The Inter-Governmental Panel on Climate Change (IPCC) gives CO_2 emissions in terms of fuel volume used. The IPCC states that burning

Carbon (mol wt 12) contained in the fuel produces Carbon Dioxide (mol wt 44) namely, 44/12 or 3.666 wt for wt. As 1.0 US gallon of gasoline contains 2421 grams of carbon, its combustion gives 2421 × 44/12 = 8.87 kg of CO_2. Similarly, 1 US gallon of diesel gives 2778 grams × 44/12 = 10.1 kg of CO_2.

For emissions in terms of distance travelled, multiply the average distance travelled by a vehicle by the emission factor for that type of vehicle (as given by the Automobile Research Association of India (ARAI)) and convert to tonne per year:

Type of Vehicle	ARAI Emission Factor (g/km)
2-wheeler (Petrol)	45.6
3-wheeler (CNG)	57.71
Motor car (Petrol)	126.5
Motor car or truck (Diesel)	166.15

For other modes of travel, the distance travelled per year × the CO_2 emission factor (given below from UK data) gives a comparative idea of CO_2 emissions in kg/year for different modes of travel:

Trains:	0.11 kg/yr
Bus:	0.09 kg/yr
Underground:	0.09 kg/yr
Ferry:	0.47 kg/yr

The average western commuter is said to burn 1290 L of fuel a year, and produce 3.4 tonne of CO_2 per year. Road transport is reported to account for fully a fifth of the UK's entire national carbon emissions totaling 33 million tonne in 2004.

Example 2: Estimate the CO_2 produced per year by a 3-wheeler using CNG and a truck using diesel oil, each traveling an average of 100 km/day.

CO_2 produced by 3-wheeler = 100 km/d × 320 d/yr × 57.8 g/km × 1/1000
= 1849 kg/yr = 1.85 tonne/yr

CO_2 produced by diesel truck
= 100 km/d × 320 d/yr × 166.15 g/km × 1/1000
= 5316 kg/yr
= 5.316 tonne/yr (nearly 3 times more).

2.4.1 Emissions from Aircraft

Emissions from aircraft travel are quite high and can be computed on the basis of fuel volume consumed and its carbon content released as carbon dioxide upon combustion. Most aircraft run on kerosene and the bulk of fuel used is kerosene. Gasoline is used only in small piston operated aircraft.

Carbon dioxide emission from air travel, as grams per passenger-km travelled basis, is the highest compared to other modes of travel, as shown below:

Mode of Travel	CO_2 Emission, g/Passenger-Km
Air travel	172
Car	131
Bus	89
Train	65

The latest AIRBUS-380 carrying 800 passengers is said to use a little less than 3L of kerosene per passenger per 100 km. Many modern aircraft use 3.5L per passenger per 100 km. Also, a modern hybrid car like the Toyota Prius uses more or less the same amount of fuel, namely, 3.8 L per 100 km when driven by 1 passenger. By comparison, in 1985 an average commercial aircraft with old technology consumed about 8L per passenger per 100 km.

A rule of thumb says that a plane and a car both emit more or less the same amount of CO_2 on per passenger per km basis, but air journeys are considerably longer than car journeys, and airplanes carry much more weight.

Aircraft generally depend on high-density fuels (ATF). Combustion of kerosene in the presence of oxygen gives carbon dioxide and water:

$$2C_{13}H_{28} + 40.O_2 = 26\ CO_2 + 28\ H_2O$$

For every 13 molecules of CO_2 produced, 14 molecules of water vapour are produced. Moreover, in case of air travel, the discharge occurs at high altitude and very cold temperature. Actually, three gases are emitted which affect global warming: CO_2, water vapour, and NO_x. Thus, in terms of warming potential the CO_2-equivalent is said to be 2.0 to 5.0 times the CO_2 released (often assumed as 2.7–3.0 times the CO_2 released).

Example 3: For covering a flight distance of 1,000 km (approx, Mumbai-Delhi) by a modern aircraft carrying 180 passengers, the fuel required @ 3.5 L, per P per 100 km is $3.5 \times 180 \times 1000/100 = 6{,}300$ L. The CO_2 emitted @ 172 g/P.km = $172 \times 180 \times 1000/1000 = 30{,}960$ kg. Assuming the warming potential as 3.0, the effective emission is 92.88 tonne of CO_2-equiv (in just a one-way journey of the aircraft).

The Directorate General of Civil Aviation, India, has decided to commission a study to determine actual carbon dioxide emissions from aircraft flying in India. As a thumb-rule, almost 10% of the fuel used in a flight is consumed during landing and take-off, both of which happen at city airports, thus adding to the city's carbon footprint. The percentage is evidently higher for short distance flights.

Food and Flower Travel

An unbelievably large amount of CO_2 is produced because food and flower items go by air from one country to another daily. Air travel is generally required to keep them fresh. We buy pineapples, fruits and flowers from tropical countries. Good for the countries, but bad for the world's environment. Buy local is the buzz word to counteract this trend. The "buy local" request is also applicable to various building materials, fittings and fixtures and a whole lot of commercially traded items. People's desire makes them seek exotic items from distant countries forgetting the twofold damage they cause to the country's local trade and the general environment. Delhi's new airport, Terminal 3, must have added millions of tonne of CO_2 to the atmosphere as a result of the heavy amount of imports reportedly made for its beautiful construction and set-up.

> One of the best innovations anyone can make to retard global warming is to develop a bio-fuel which can be safely used by aircraft all over the world. Jatropha is a carbon-neutral fuel, but is Jatropha a fuel of high enough density to serve as an aircraft fuel? Some actual tests seem to say so! More tests are awaited!

A whole new industry would develop with the use of carbon-neutral biofuels and the global warming problem immensely reduced in the bargain. Till that time, "buy local" and "travel less by air" are requested. See Section 11.2 for more ways to reduce emissions.

2.5 IMPORTANT ALTERNATIVE SOURCES OF ENERGY

Fortunately, there are several important alternative sources of energy available to us – some renewable, some long lasting, most free from CO_2 production, and some totally unexplored by us at present. These sources are all listed below to show that a large number of alternative forms are already available for exploitation. They are grouped according to their readiness for application.

Alternative sources already in use

- Hydro-power
- Agricultural biomass and Community wastes
- Municipal solid wastes, and Waste-to-energy plants
- Coal gas, natural gas, LNG, CNG, LPG
- Nuclear energy

Alternative sources ready for use

- Wind
- Solar
- Hydrid (both wind and solar together)
- Biofuels from oil-bearing plants,

Alternative sources in development stage

- Shale gas
- Wave energy
- Geothermal energy
- Satellite-solar

As shown above, some methods are already in use and we have experience in using them. Some are ready for use. Others need R & D work before they can be used.

> **Observation:** The principal advantage in the use of alternative sources of energy is that except in the case of bio-fuels, no CO_2 is released to the environment as a result of combustion. As we see later the bio-fuels release some carbon but they are essentially considered "carbon-neutral" because the carbon was drawn from the atmosphere earlier, so there is no net addition of carbon!

2.5.1 Green Energy

A term often used in referring to alternative forms of energy which do not produce carbon upon combustion is GREEN ENERGY. Green is not a colour, but perhaps an acronym for **G**rowth with **R**esources, **E**nvironment **E**nhancement and **N**ature.

The technology is there for use of alternative sources of energy and a few installations are in use. The price is yet a little high for general use. However, in certain specific situations even the price becomes favourable when all things are considered such as cost of long lead lines, transport cost of diesel, etc. Nuclear energy appears to be already affordable in cost per unit, but needs much greater public support as far as safety and security are concerned. Greater use of bio-fuels for blending with fossil fuels is also feasible but needs more countrywide marketing effort. Wave and Geothermal energy need more technical and scientific study to make them applicable under Indian conditions. Research and development are needed.

The Chinese have a very apt and useful proverb which says: *Drink Water, Think Source!* We have a somewhat similar request to make when using fuels:

THINK RESOURCES, THINK EMISSIONS!

- **With fossil fuels (coal, oil, petroleum products):** Both resources and emissions are a big problem.
- **With natural gas, LNG, CNG, LPG:** Both resources and emissions are a lesser problem. Hopefully, shale gas will be an addition to this group.
- **With agricultural and community wastes (sewage):** Limited supply or resource is a problem. Emission is not a problem.
- **With Hydro Power:** No emission problem. But, resource is a problem (limited and uncertain rainfall).
- **Nuclear Power:** No emission problem. But safety/security and land availability problem.

2.6 ALTERNATIVE SOURCES ALREADY IN USE

Let us first see the alternative sources of energy which are already in use and with which we are familiar.

2.6.1 Hydropower Projects

In a monsoon country like India, large hydropower projects have been built since ages and are considered so important that a separate ministry is devoted to handle the subject. Bhutan also has several hydropower projects because of its mountainous terrain, but the electricity so produced is solely for the benefit of its neighbor, India. US, Germany, China and India rank high in the world.

In fact, hydropower projects are preferred in modern times, wherever possible, because no CO_2 is released in generating electricity therefrom, neither does the water go to waste after use in the turbines.

In hydropower projects rainwater is harvested in a natural or man-made reservoir located on a high terrain and released as desired to drive turbines located in a valley below to generate electricity, thus converting the potential energy of stored water into electrical energy. After use, the water is released to a river downstream for supply to towns further downstream, or the flowing water can provide energy, if desired, to operate a wheel or hydraulic ram to pump up some small quantity of water.

Mumbai has been served for many years since early 1900s by a large and dependable hydro-electric scheme (Tata Hydro at Bhivpuri and at Lonavla, near Mumbai) with the monsoon filling up its reservoirs and renewing their life each year on time. Tatas are now toying with the idea of placing floating solar PV panels on the lake water surface so as to make double use of the lake land. There are several other examples of commendable hydro-electric and irrigation projects in India.

Double Use of Land

Tatas (along with an Australian firm) are now reported to be exploring the use of a new type of hybrid system in which a flexible solar energy photovoltaic system is placed on their existing lake's water surface at Lonavla so as to get additional electricity from solar sources while continuing to benefit from the earlier hydro-electric source below. This avoids the necessity of buying additional land (which is difficult to get in India). If it proves successful, it will work wonders for the environment and costs can come down still further because, firstly, they will continue to get electricity from the stored water, and secondly, they will get additional electricity from solar sources, thus making double use of the land.

Energy of Flowing Water

After the lake water operates the turbines, it is released downstream into a river where the energy of flowing water can be further harnessed to rotate a wheel or operate a hydraulic ram, if desired. Such devices have been used since time immemorial, though many are not to be seen nowadays. Water is also picked up from the river downstream for supply to adjoining towns. Thus, the lake water serves more purposes.

Small hydropower projects upto 25 MW capacity are reported to remain a problem because most of the streams and rivulets in India are not perennial. Any attempt at creating small dams on such streams causes objections from the downstream farmers and also often leads to deforestation. Nonetheless, about 2,700 MW of electricity is produced in this manner at present. Pumps (operated with electricity or diesel) have been used.

Agricultural Biomass and Community Wastes

Agricultural biomass (crop residues, sugarcane bagasse, leaves, twigs, grass, etc.) and community wastes (sewage, industrial wastewaters from the food industry, etc.) can all be a good source of biogas. Wastes like sewage and other wastewaters rich in organics are sent to municipal wastewater treatment plants. The reader is referred to Chapter 7 for further discussion.

Some specific wastes from market places and industries high in organic content may be treated in industry-owned "Upflow Anaerobic Sludge Blanket" (UASB) plants or other "waste-to-energy" plants for anaerobic decomposition. These are also discussed in Chapter 7.

Municipal Solid Wastes Disposal

In India, as we shall see in Chapter 9, the conventional system of handling solid wastes has been to use petrol or diesel driven vehicles for collection of the waste from roadside bins or from houses, followed by transport to a low-lying dump-site for disposal by dumping. The waste is left there to undergo self-stabilization and the gases produced are allowed to escape to the open atmosphere. In some of the larger cities, the people may be encouraged to segregate the waste so that salvageable material is picked up separately by rag pickers and recycled into the market. The collection system in large cities may also benefit from use of waste compression trucks and transfer stations so as to be able to carry more waste per trip. Ultimately, all solid waste of the city goes to the dump or landfill site for final disposal.

The collection vehicles They presently use diesel oil mostly. For India it is an imported item. These vehicles should either run on biogas or CNG or be converted to EV (with electricity generated by solar panels or UASB-type anaerobic systems with dual-fuel engines in tow).

Solid wastes disposal Modern trend in case of larger cities is to go for sanitary landfill or aerobic composting instead of plain dumping of waste in low-lying areas although composting is relatively more expensive than dumping because the compost has to be transported to the agricultural fields after it is ready. Where more methodical and expensive landfilling of waste is done, a biogas collection network may be installed in the landfill which would help to collect the biogas for sale as fuel rather than its release to the atmosphere. It would also help meet costs. Gas collection and utilization is also favoured these days because of global warming considerations and the income it generates for the local body.

Incineration of the solid waste instead of dumping or landfill is done mostly in Western countries, not yet in India where the waste has a relatively low calorific value. Across Europe there are over 400 plants of the "waste-to-heat" type. Such plants use the city's solid wastes as fuel instead of oil or other fossil fuels and have efficient waste gas cleaning devices incorporated in them so as to remove mercury, dioxine, CO_2, etc., besides smoke, soot and dust. Their excess heat is piped for use in adjoining homes or for producing electricity. Denmark, Germany and Netherlands seem to benefit most from such plants. They save land too.

Delhi is reported to be getting its first "waste-to-energy" plant based on 2,000 tonne of the city's solid waste and capable of producing 16 MW of electric power in the first phase. Even then, the plant to be established by Jindal Ecopolis at a cost of ₹ 200 crore, will consume only $1/3^{rd}$ of the city's solid waste.

2.6.2 Coal Gas

Coal deposits are substantial in India and coal gas was once produced in large quantity. Coal gas is also referred to as town gas and in the past was the principal fuel for domestic heating and cooking. It has now tended to be replaced by natural gas. Coal was also used in the past to generate "Producer Gas" to drive solid wastes collection vehicles (especially in the War time when imported petrol was scarce but coal was locally available). However, it has tended to lose favour as it is based on a polluting, fossil fuel like coal.

2.6.3 Natural Gas Deposits

Natural gas deposits have been found in abundance in the Qatar region of the Middle East, as well as in Russia and in the North Sea above Britain. The gas is composed mainly (>90%) of methane. It is odourless, non-toxic and non-corrosive. It liquefies at atmospheric pressure when cooled to –160°C. Qatar has invested heavily in refrigerated ships to transport Liquid Natural Gas (LNG) to other countries, whereas Europe receives its supplies through pipes laid from Russia. Natural gas has been extensively used in some cities of India owing to its relatively non-polluting nature and ease of distribution by pipelines.

India has also been fortunate to recently find enormous *gas resources* which will no doubt be used extensively in the years to come. The gas deposits found in the Krishna-Godavri basin will last for many years to come. They will, in fact, slow down to some extent the development of wind and solar energy unless the latter are made mandatory or some tax benefits are given. Moreover, gas turbines are reported to have a 60% efficiency of conversion to power, which is only next to nuclear power. Gas resources seem to have an excellent future in India.

Compressed Natural Gas (CNG) is the same gas supplied at high pressure in small cylinders and is used both in households as well as in transport. Mumbai's 55,000 old taxis working on diesel or petrol have switched over to CNG as a means of controlling air pollution from their tail-pipes. Its attractive pricing has no doubt helped to ease its ready adoption by the taxis. The use of CNG is also likely to increase in the future as populations increase and as the people find it convenient and safe to use it increasingly for transport purposes.

Liquid Petroleum Gas (LPG) is different from LNG. LPG is almost all (95%) Propane, and some Butane. No cooling is needed but under heavy pressure it liquefies. It is also sometimes used in transport.

2.6.4 Nuclear Energy

In India, at the present time, development of nuclear energy (though not strictly renewable) is being considered although the Chernobyl and recent Japan disasters are much feared. Its efficiency of conversion to power is high, upwards of 70%, and it can reduce India's power shortage without producing any CO_2. India's experience with nuclear energy in the past has been very satisfactory. India already has a few nuclear power plants. But, their safety aspects are facing fresh doubts. Additional safety precautions, however, can be built into new plants from experiences gained elsewhere in the world.

Besides safety aspects, the large land requirement of nuclear energy plants is a severe deterrent. A large chunk of land has to be left vacant surrounding a nuclear plant from safety considerations. An additional chunk of land around this area has to be left vacant for controlled activities and controlled habitation. And yet another chunk of land has to be provided nearby for holding the nuclear waste for long periods. Its land requirement seems to be even more than that for solar concentrators. Where is so much land to come from?

Land availability becomes the most severe problem in undertaking a nuclear project. The recent Jaitapur Nuclear project has high-lighted these difficulties and showed that people are not so ready to voluntarily part with their existing land. Land acquisition policy of government has to be reviewed and made more people-friendly, if it has to succeed.

Besides safety and land, another constraint is water. Large volumes of water are needed as cooling waters in the reactors. Their location is therefore often limited to coastal areas with easily accessible sea water. A danger which comes with seaside location is the danger of Tsunami action. Similarly, when located inland near large and perennial rivers, earthquake zones have to be avoided.

Another problem is the long time period required for getting international permits for nuclear plants and their construction (about 8–10 years). Thus, investments have very long pay-back periods. Nonetheless, some countries like France have 80% of their electric power generated from nuclear sources.

> One of the severe problems in developing nuclear energy is its enormous land requirement, (apart from safety).

2.7 ALTERNATIVE SOURCES OF ENERGY READY TO USE

The principal renewable sources of energy available for use in India up to end of March 2010 are according to the Ministry of New and Renewable Energy (MNRE) the following:

Wind Power (India is said to be 4th largest)	11,807 MW
Solar Power	12 MW
Solar PV street lighting systems	88,300 nos

Solar PV pumps	7,300 nos
Biomass Power (non-bagasse, agro residues)	1,100 MW
Waste-to-energy	110
Solar water heating, collector area	3.53 Mill sq m
Solar lanterns, pieces	8.0 Mill pieces
Family-type Biogas plants	4.2 mill units
Biomass cooking stoves	35.2 mill units

Yet with all these activities, renewables constitute less than 3% of all energy consumed in India at present. Our target is to get to around 20% by 2020. Cost is one reason. Solar energy for example, costs more at present than energy produced from oil or coal. The whole scale of activity is so small at present that unit costs are high and we also have to depend on imports. Subsidies have to be given to make things attractive at present.

> Some argue that if subsidies have to be given, let them not be given on imported items like oil; let them be given for energy sources where we remain independent of any imports and develop our own resources.

Let us look more closely at wind and solar energy first as renewable sources of energy for domestic and industrial purposes. For general usage, their costs are high at present and some fiscal incentives are needed to make them attractive. However, even now their costs are "affordable" under certain circumstances discussed later. Costs will surely come down in near future.

2.7.1 Wind Energy

A wind energy plant on average needs an investment of ₹ 5.5 crore per MW, compared to ₹ 4 crore for a thermal (oil-based) plant and ₹ 5 crore for a hydropower plant, today. Wind energy is likely to become cost competitive very soon.

To promote greater use of wind energy in India, many more wind mills, wind-turbines, and wind farms need to be set up to generate electric power where wind is found to be blowing reliably. The key word is "reliable". In India, wind is generally strong during the monsoon months but often very weak thereafter.

Several steps, such as the following, are necessary to take in order to encourage the general public to install and use modern, better quality, low-friction, wind-operated systems:

Firstly, locate geographic areas in India where wind farms can be set up assuredly People do not know where wind is strong and where it is weak. In certain locations, e.g., coastal areas, hilly areas and other specific locations where what is known as "wind-tunneling" effect occurs, all make excellent sites for installing wind turbines.

Everybody knows that wind does not keep blowing all the time. It stops after a while. In most countries, it generates power for only 20 percent of the time, and either needs storage batteries to tide over other times or needs hybrid installations including solar systems along with wind (see later). Currently, ideal average wind speeds are reported to be between 7 and 9 meters per second (15–20 miles per hour) for utility-scale wind turbines suitable for commercial applications.

Power obtainable from wind is a function of the cube of the speed. A wind speed increase from 6.5 to 9 m/sec would double the power output, no small benefit. India is unfortunately reported to be in a region of low wind speeds barring the monsoon months. Strong winds blow during the monsoon season only. (The Ministry has estimated a total potential of 45,000 MW only from wind though new research would tend towards more optimistic predictions).

This kind of wind speed is best achieved by going above trees and buildings, preferably on building terraces. One advantage of locating on terraces is additional land is not needed. However, structural stability has to be assured in each case and this is best done early on at the building design stage rather than afterwards.

Secondly, give fiscal incentives to promote more installations Already, wind power acquisitions are seen as a tax-saving device since accelerated depreciation at 80% is allowed in the first year itself. Another incentive can be offered in form of enabling reverse selling of surplus electric power to national power grids so that those who generate power can also earn some money from it. Reverse selling is in a way encouraging public-private-partnerships to increase number and size of installations. This topic is further discussed under solar energy.

Thirdly, where land is a problem, consider sea-based installations The Norwegian energy giant "StatoilHydro" has successfully built the world's first full-scale floating wind farm on the open sea as an experiment to harvest electricity in an environment of strong, low-level winds and fewer space restrictions than are found on land. US and Germany have also approved construction of the first offshore wind farms in Massachusetts and in the North Sea, respectively. This seems to obviate the necessity of large-scale land acquisition which is today becoming a severe problem in India.

Incidentally, the following requirements for wind systems may sound somewhat discouraging but are not so in most cases:

Before wind turbines or wind farms can be installed in certain places, one may need permits from appropriate authorities involving submission of (i) a noise study (ii) an avian study to determine potential impacts on birds and bats, and (iii) an aviation administration study to determine if the wind turbine might interfere with aviation radar. Their clearances must first be obtained.

2.7.2 India's Efforts

A newly formed company called Green Infra Ltd. (GIL) sponsored by IDFC Private Equity, has bought up a 60 MW wind farm in Karnataka and another 40 MW installation in Maharashtra from British Petroleum's BP Energy. GIL is also looking to buy 250 MW wind energy capacity from DLF and eventually reach a total of 500 MW capacity in 3 years. ONGC has set up a 50 MW wind farm project in Kutch District at Bhuj, Gujarat, using 1.5 MW machines installed on 78 m high towers. The technology used is said to be nearly maintenance-free and saves the organization nearly ₹ 30 crore per year in electricity charges.

India is reported to be ranking fifth in total installed capacity with 9,645 MW of wind power installed by the end of 2008. A few foreign-based companies are setting up shops in India to benefit from lower costs and lack of conventional electric grids in small and medium-sized cities. In Europe, installation costs are said to be around $ 2.20 for every megawatt of off-shore wind built. In India, similar installations are expected to cost 30–40% less. Where much power is needed (e.g., for an industry or a small township) a single wind turbine may not suffice and a wind farm may have to be set up. In such cases, enough land may need to be acquired for the purpose. Land acquisition may pose a severe problem as explained later.

Wind energy systems have been provided in India by Indian as well as foreign companies. Some Indian companies have also provided wind energy systems in other countries. An Indian firm called "Suzlon" is today a world giant in providing large-sized wind energy systems both in India and overseas to which it exports.

2.7.3 Solar Energy Systems

As a tropical country which receives much sunlight, India almost tops the world (along with California and Hawaii in the US) in its *potential* for developing solar energy systems. Spain, Italy, Australia, and China come next on the list. However, in actual implementation Germany and Spain top the list at present.

Solar energy systems consist of the following:

- Solar thermal (solar cookers, solar water heaters)
- Solar photovoltaic (stand-alone or grid systems for producing electricity)
- Solar concentrators (for larger installations producing electricity)

Indian manufacturers are presently offering only (1) small solar items such as solar activated lanterns and cookers, (2) solar water heaters for bathrooms, and (3) photovoltaic panels manufactured in India. Concentrated solar power systems for larger habitations exist in Spain and USA, but have not yet been installed in India. Many Indian companies ranging from Tatas, Birlas, Jindals and many more are now jumping on the solar bandwagon! JSW Green Energy an Indian Company belonging to the JSW group is planning to use in India a unique lightweight and flexible photovoltaic device developed by an American group.

Solar Cookers

Solar cookers have been in use since a long time. They have very useful application in case of schools, hostels, etc., where meals have to be prepared usually in large numbers. They are also useful for religious centres where large crowds visit on certain days and meals have to be prepared for them. Their use makes cooking cheaper and more hygienic and also saves production of CO_2 which would be emitted from use of other cooking media like LPG, wood, charcoal, etc. There are cases where some organizations have benefited from Certified Emission Reductions (CERs) and Carbon Credits have been profitably sold on the world market.

Solar Water Heaters

Solar water heaters are equally popular for bathrooms and kitchens where hot water is required in relatively large volumes for hostels, hotels, hospitals, guest houses, industry canteens and such purposes. Such systems have proved popular with large-scale users for whom payback periods as short as 2 years, have been noted. Several installations of this type have sprung up in different parts of the world. Their use is particularly economic where large volumes of hot water are continuously needed each day, such as in hotels, hostels, hospitals, commercial kitchens, etc. Separate plumbing systems may be needed from the roof to the bathrooms to benefit from solar hot water systems. For smaller systems for individual families, the payback periods are much longer and governmental subsidies may be needed to make them attractive.

Their use could, and should be made mandatory in case of larger installations. A solar hot water system that gives 125 liters per day comes for about ₹ 20,000/- only, and saves a lot of fossil fuel.

Solar (PV) Lanterns

Solar activated lanterns, (TERI-type) are the simplest and smallest type of devices which use solar energy. They require the use of small handy photo-voltaic panels and have relatively short payback periods. They constitute an excellent answer to the absence of conventional electrical power for lights in rural and semi-rural areas of India. Only an initial investment in the unit has to be made. Thereafter, solar radiation generates the electric power for free. But the power needs to be stored in a battery. The running cost consists of replacement of the battery every 2–3 years. By the year 2022, India expects to install 20 million solar lights which would result in a saving of about 1 billion litres of kerosene every year.

A company based in Delhi has developed a solar powered unit using an LED-based lamp that is waterproof, portable and runs for 12 hours on the stored power. It is likely to be an attractive replacement for the conventional kerosene lamp.

Solar (PV) Panel, Stand-alone Systems

Two types of PV installations with panels are normally made: stand-alone systems and grid connected systems.

Stand-alone sources of renewable energy find favour in rural and urban areas of India for generating electricity at the spot where it is required to operate one of the following:

- a tube-well pump, or
- an intermediate station for mobile phones, or
- small water treatment units to remove hardness, arsenic or fluorides and such other impurity from locally available drinking water.
- small wastewater treatment units like aerated lagoons with floating aerators.

The real cost of generation of diesel power is reported to come to around ₹ 17 per kWh if we add the subsidy component. This cost is more or less the same for solar PV power. Advantage with solar power is that we don't need to import oil and transport it to the site.

A good example of stand-alone systems are the telephone poles provided along the Mumbai-Pune expressway, each carrying its own PV panel on top to generate electricity to run the telephone lines.

According to Prof Bannerjee, Head of Energy Systems Engineering, IIT Bombay, the capacity of solar panels varies widely starting from 2 kW to about 200 to 250 kW. Prof Bannerjee, refers to them as modular units, much like lego sets which can be readily put together. If you intend to set up a 1 kW installation, you would need the panels with their mounting structure, a battery to store the power, an inverter unit to convert from DC to AC, wires, cables, etc., which along with labour would cost about ₹ 3 lakh upfront. The batteries, costing about ₹ 15,000 would need to be replaced every 3–5 years and one would end up spending ₹ 12–15 per kWh. The PV panels often carry a 25-year guarantee but are likely to last even longer.

A company has also stepped in to meet the growing shortage of drinking water by designing a solar system to operate a water generator that converts water from the atmosphere (humidity) into 500–1000 L of fresh and clean drinking water. They have also designed a module with a solar panel suitable for mobile towers. These units replace the diesel generator sets which currently serve as back-up for the towers. The Company (Surana Ventures Ltd.) is providing 2,000 stand-alone type Tel kiosks and ATMs in New Delhi for the Commonwealth Games.

The use of floating solar panels is also being suggested with aerated lagoons so as to produce electricity at site for operating the aerators in un-electrified regions, without having to buy extra land for placing the panels.

Solar (PV), Rid-connected Systems

Grid connected systems with solar panels are generally fitted on roofs in urban areas. Thus, no additional land is required as roofs are used. This can be a great advantage in crowded urban areas.

Ideally, all such systems should be in full sun from 9 am to 3 pm in mid-winter.

Installing PV systems is reported to be expensive at present. Costs will hopefully come down from ₹ 15–16 per unit without subsidy (₹ 9 per unit reported presently with subsidy). It is interesting to note what an eminent industrialist states that companies presently spend as much as ₹ 30 per unit with a "genset". They could as well save some money by switching over to solar PV panels (without any subsidy) and what is more solar energy would be entirely sustainable and would cut down on cost of transporting fossil fuels.

> Some of the installations that have come up presently in India in spite of high costs are because of the desire to break free from long hours of load-shedding which would otherwise have to be faced by the industries, especially in smaller Indian towns. Load-shedding is neither good for industries nor for wastewater treatment.

A conventional-type system capable of giving about 1 kW of energy would need a capital investment of about ₹ 3 lakh upfront. Sometimes, the installation costs may increase by another ₹ 1 lakh if difficult conditions are encountered at site. As more installations come up in India, costs would come down and pay-back periods get reduced. Tax incentives and subsidies would also help.

An advantage of PV systems, however, is that long transmission lines are not necessary, nor is much land required to install them. They are generally fitted on to the building roofs. More and more solar modules are now being fabricated as building materials so that they can be integrated into the building's fabric by architects. They include solar roof tiles, wall materials and semi-transparent roof material for atriums and skylights. These are destined to increase usage in the near future.

2.7.4 Reverse Feed-in

An attractive inducement for solar installations is payment for "feedback" of electricity from private individuals into the public grid. A kind of reverse feed-in. It would encourage private capital to come in if payment is made at a rate that would make such feedback profitable. A recent governmental ruling in India requires all electric supply companies and electric boards to have at least 5–7% of their energy supplied from renewable sources, and pay ₹ 18.50/- per kWh for captive feed-in systems. This makes feed-in an attractive proposition. (In Germany the stipulation is 15% of energy must be from renewable sources). This should give a new push to installation of renewable energy systems for feed-in into the existing grids on private initiative in urban areas of power short cities.

The Indian Oil Company (IOC) is reported to be considering setting up in association with Bharat Heavy Electronics Ltd (BHEL) a solar power plant worth about ₹ 100 crore in Barmer, Rajasthan, based on the PV panel technology. The Government hopes that by 2022, solar power installation in India will reach 20,000 MW. It also hopes that 9,000 villages will benefit from soft loan schemes which will be refinanced by the Indian Renewable Development Agency Limited. Rooftop solar power will be promoted by providing incentives for self-use. To enhance human resources, the technology institutes are being asked to develop suitable training courses.

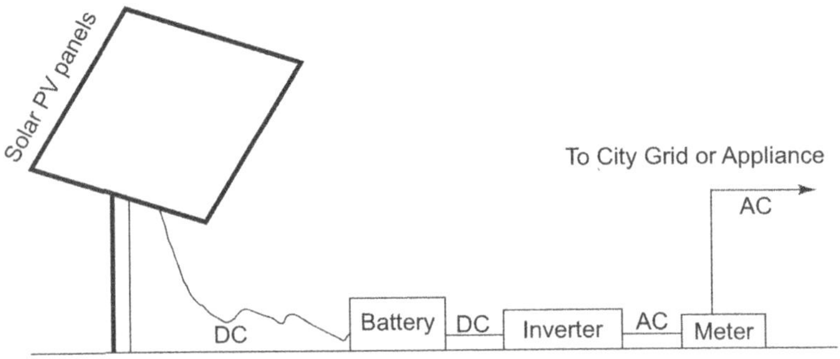

Fig. 2.1 *Use of Solar PV Panels to generate electricity for feed-in into city grid*

2.7.5 National Targets

The Jawaharlal Nehru National Solar Mission targets 1100 MW grid solar power, 7 million sq meters solar collectors and 200 MW off-grid solar applications in the first phase by 2013, and 20,000 MW of grid solar power, 20 million sq meters solar collectors and 2000 MW off-grid solar applications by the year 2022. There are several off-grid applications which are already commercially viable or near viable where rapid scale-up is possible, with resulting reduction in imports of fossil fuels.

Observation: The Indian authorities are hopeful that the target of 20,000 MW of solar generating capacity by the year 2022 will be met even earlier than that year and India will become a global exporter of solar power systems.

Under the National Solar Mission, three major initiatives are proposed to be undertaken: (1) Create volumes to encourage large-scale manufacture, (2) announce a long-term policy to purchase power from renewable sources, and (3) support R & D.

A Solar Energy Society of India (SESI) has been formed to promote dissemination of information, co-operate with other research bodies, NGOs, etc., and generally to promote R & D in the subject, and further its applications. According to SESI, most parts of India receive solar energy of 4–7 kWh per sq meter per day which signifies about 5,000 trillion kWh per year of energy over the whole of India.

According to guidelines issued by the Central Electricity Regulatory Commission (CERC), India, investments in renewable energy plants such as solar, wind and biomass, will get a pre-tax return of 19% per year for the first 10 years and 24% per year from 11th year onwards.

2.7.6 Concentrated Solar Power Systems

While efficiency of PV panels is only 10–15%, it is relatively higher at 24% for Concentrated Solar systems using parabolic mirrors.

Large-sized plants for supplying electricity to urban areas and large industries are presently coming up in a few parts of the world to demonstrate various new technologies. Experience has shown through the development of special large-sized units called *"solar concentrators"* for power generation in Spain, US and elsewhere that it is possible to generate power from solar energy on a large scale but there are still a few problems yet involved in development of solar power for cities and large consumers.

Remember solar energy is plentiful in nature but available on the earth in "dilute" form. Hence more land is needed (about 10,000 acres for 1,000 MW plant, the land alone costing around ₹ 100–200 crore!). Cheaper land would be available in an area like the Rajasthan desert. Also, from the experience of Spanish installations, large parabolic mirrors shaped as ultra-long troughs are needed to concentrate sunlight on to a long tube of molten salts that store heat. The heat energy is stored in molten salt (actually a mixture of two fertilizers, sodium nitrate and potassium nitrate). This heat is then used in a conventional thermal power plant to give electricity at about four rupees per unit. It is hoped that with further research and improvements in future the Concentrated Solar Power (CSP) system costs may become competitive with conventional power and larger-sized units could be developed. There are already signs that costs are sure to come down for both CSP and photovoltaic panels.

Land availability may, however, become a problem in India as experience in Singur and Nandigram, West Bengal, has shown. Even though desert lands can be brought into use without affecting agriculture, often, even what is classified as wasteland by Government, is considered valuable by villagers who use the land for grazing, collecting minor timber, and as transport corridors. Villagers have also spurned joint ventures with corporate bodies out of mistrust. Hence, land can become a serious problem in India.

Moreover, water is also needed to wash off dust from the solar mirrors. This water may be scarce in wasteland areas and need expensive reuse techniques to make it possible. Use of solar energy is thus not so simple a problem as it may seem. Another problem is how to store heat produced during the day for later use. First commercial solar plant with heat storage opened in 2008 in Guadix, Spain, east of Granada. During the day the molten salts are heated, while in the evening it gives back heat to make more steam. A Spanish company is now building a plant in Arizona, USA, for Phoenix and Tucson which will generate 280 MW in 2012.

Germany and Spain have invested heavily in solar power. Germany has greater than 5 Gigawatts capacity and is the largest producer of photovoltaic power. Electricity generated from solar energy (in 2006) in gigawatt-hours is reported to be: Germany 2220, US 565, S. Africa 532, Spain 125, China 105, Australia 31, S. Korea 31, Italy, Netherlands, Switzerland, 20–30 each, Canada 21, India 19, Mexico 10.

In May 2011, a news item states that the Maharashtra State Cabinet approved the setting up of a 150 MW solar power project (touted as the largest in the world) at Shivaji Nagar in Dhule District. It is expected to be commissioned by March 2012. The Maharashtra State Power Generation Company (Mahagenco) has already commissioned two other solar energy projects in Chandrapur. The cost of this project is expected to be about 2,000 crore of which 20% will be borne by the State and 80% will be loaned by KFW, a German financial institution. Electricity from the new plant will be sold at ₹ 17.90 per unit against its expected generation cost of ₹ 12 per unit.

A fantastic project is hopefully being conceived using the large and barren Sahara Desert area in the north of Africa to develop a concentrated solar power device to generate enough electric power to meet the entire energy needs of Europe! The energy generated would also enable desalination of sea water needed for humidification to grow crops in greenhouses located in the Sahara Desert. It is an excellent example of synergy between CSP, seawater and greenhouses. Land is available, solar energy is available, generation of power is possible, but the conveyance and distribution of power from the desert to various European countries would pose enormous problems and involve large costs. Of course, no greenhouse gases would be produced. In India, the deserts of Rajasthan and Thar might hopefully be developed one day on similar lines as the Sahara though several other areas are also likely to be considered.

2.7.7 Satellite-based Solar Power Systems

The land problem is the most serious problem in India both for solar power and nuclear power development. In the case of solar power, the problem could perhaps be avoided to some extent by the recently planned development of Satellite-based Solar Power Systems under which solar collector panels would be placed in a geo-synchronous orbit in space (to benefit from 24-hour sunlight) and solar power beamed to earth in laser or other form for subsequent conversion to useful electricity. These technologies are not yet fully ready for commercial exploitation but are mentioned to show the lines on which thinking is going on. This is where the future of solar power lies. A 24-hour supply and no land problem! Let us hope organizations like ISRO will take up the challenge someday soon.

2.7.8 Hybrid (Wind-Solar Integrated) Systems for Buildings, Transport and Industries

Hybrid systems are installed where solar and wind-operated systems are both used to complement one another. Solar systems would not be available at night or in cloudy weather, while wind-driven systems would not work at no-wind times. Thus, the two could be made to complement one another to keep power supply available for longer time of day and night. Of course, storage batteries would be needed.

> "Green" buildings are provided with better planning and design of the structure architecturally and engineering-wise, so as to reduce the building's carbon footprint. Second generation "green" buildings would in addition to better engineering and architecture, also generate some electric power without adding to its carbon footprint (see Section 11.3).

2.7.9 Second Generation Green Building

Mention must be made of a building in north Pune ("Orange County") which I call "second generation" green building, where wind-mills and solar PV panels have been installed to generate enough electricity to operate the building's lifts and some common services besides giving free "green" energy to all its 36 flat owners. As said earlier, this becomes a case of Public-Private-Partnership which is welcome because Government is perennially short of funds and welcomes the input of private funds for electricity. The builder benefits from a valuable selling point to attract flat buyers who would prefer free electricity rather than a gymnasium or some other attraction thrown in by the builder when selling the flat.

2.7.10 Use of Electrical Vehicles for Transport: PV Solar Panel + Fuel Cell

A unique arrangement is feasible thanks to an MIT researcher having developed a cheap self-generating catalyst recently, to run an electrolyzer which is hooked up with a PV solar panel (see Fig. 2.2). The PV solar panel produces electricity during the daytime to meet domestic needs. After meeting the daytime needs of the building, surplus electricity is taken to an electrolyzer which generates H and O from water and stores it for recombining later in the dark hours of the evening in a fuel cell to generate electricity and give water as a bye-product. The electricity could be used for domestic purposes in the night and also to recharge the batteries of an electric vehicle parked in the garage. This would make it possible to use Electric Vehicles.

If a fuel cell is not available, the surplus electric energy may be stored in a set of batteries for use during non-sunlight hours as mentioned above.

> This possibility would make use of Electric Vehicles (EVs) become a reality sooner than we think. Use the EV all day and let it recharge at night in your own garage in the building where you stay. Traffic pollution as well as car maintenance would be eased, and no petrol need be purchased at all! This is too good to be true in the age of vested interests of oil companies!

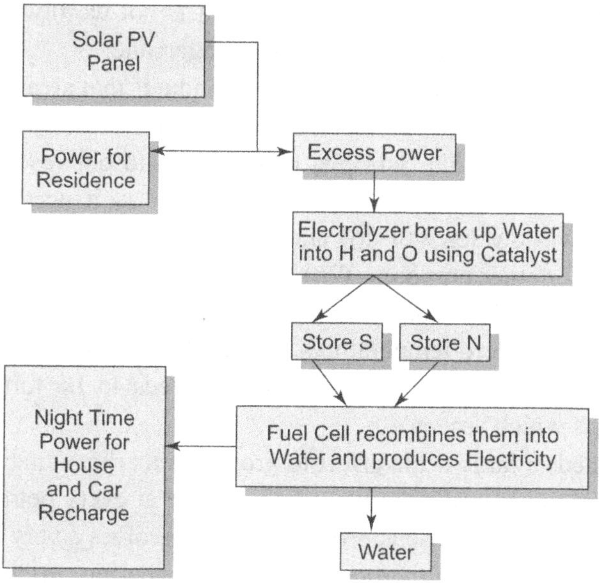

Fig. 2.2 *Use of Solar PV panels + Fuel cells to provide continuous electricity to a house and car*

2.7.11 Carbon-neutral Bio-fuels for Transport

There are two prominent types of bio-fuels which can be used in the transport sector, both obtained from agricultural sources:

1. Cultivation of "Jetropha" plant (an oil-bearing plant) whose oil can be used as an additive with petrol or diesel.
2. Another bio-fuel is obtained through chemical conversion of agricultural products such as corn or sugarcane to give "Ethanol".

Both Jetropha and "Ethanol" can be used as additives or as alternate fuel. Further, Brazil has shown that a mix of 5–10% bio-fuel along with regular petrol can be used for cars without any changes to their mechanism. At 5% blending, a country like India is estimated to need about 600 million litres per year and at 10% blending it would need double this quantity.

However, use of Ethanol has been somewhat controversial. Its use appears desirable where *sugarcane* is the source from which Ethanol is made as the energy balance is of the order of 8:1. In colder countries where Ethanol is made from *corn,* the energy balance is considered to be so unfavourable that when one considers all inputs like water, fertilizers, gasoline and transport for cultivating corn, it is not worthwhile to make Ethanol. India and Brazil fortunately grow sugarcane.

The problem in a populous country like India is that area available for food cultivation and water cannot be diverted to make way for bio-fuels instead of food. Here, the Jatropha plant comes to our rescue. Jatropha is regarded as a better alternative to sugarcane as it needs much less water (even wastewater will do) and can be grown on land considered "wasteland" for other purposes. With diesel engines, Jatropha can even be used unblended on 100% basis. Jetropha has performed well even on high mountain and sub-zero temperatures. The use of Jatropha oil for fuelling aircraft also has tempting possibilities but needs to be further proved through trials which are on.

India needs a large-scale effort to grow and market suitable bio-fuels like Jetropha and to make them easily available at every petrol station in the country so that their use is promoted. This involves a very large-scale, countrywide effort at cultivation, storage, distribution, pricing, etc., for the bio-fuel on the same lines as is being done by government agencies for marketing petrol and diesel. It has employment potential at village, small-town and large city levels.

The Indian Oil Corporation (IOC) is reputed to have made a welcome beginning along with the Chhattisgarh Renewable Energy Development Agency in cultivating Jatropha on 600 hectare waste land in Madhya Pradesh. It is hoped Government will continue to show consistency in its policies for promoting use of bio-additives irrespective of the world price of oil.

Carbon-neutral Bio-fuels

It is necessary to clarify that both fossil fuels and bio-fuels produce CO_2 when burnt. In the case of fossil fuels like coal or oil, the carbon sequestered in it for millions of years is released when burnt and gives an extra load of carbon on the environment which can lead to global warming. In case of bio-fuels, CO_2 is no doubt released from the carbon contained in the oil, but it was extracted earlier from the atmosphere through photosynthesis which generated the plant in the first place. Hence, bio-fuels are considered as "carbon-neutral".

> The growth of bio-fuel plants absorbs CO_2 through photosynthesis while their burning a little later just replaces the CO_2 back in the atmosphere. So they are effectively called "carbon-neutral" fuels.

2.8 ALTERNATIVE SOURCES OF ENERGY NEEDING DEVELOPMENT

2.8.1 Shale Gas

Shale is a sedimentary rock that can hold gas and oil. Shale gas is a natural gas found in rock formations and requires specialized production techniques. The USA first produced shale gas in 1989 and is reported to have later developed a new and easier method of extracting gas from shale, a common sedimentary rock found across the world. Shale gas exploration entails what is called "fracking" (hydraulic fracturing) or vertical drilling followed by horizontal drilling. Shale gas is found in tight reservoirs that have poor permeability. Shale gas requires large number of wells as the volume obtained per well is small. In 2009, shale gas accounted for as much as 13% of total energy consumption of US. Almost all states in US have reserves of shale gas (totaling 24.4 trillion cu meters in US). China is reputed to have the largest reserves of shale gas (36 trillion cu meters) and are getting into the act now. Reserves have been found in Argentina, Mexico, South Africa, Australia, Canada, Brazil, etc. Potentially, every rock formation can have shale gas.

India also has massive shale deposits and thus there is good scope for use of this gas in future. Reliance India as well as ONGC are reported to be tying up with US companies in order to benefit from their exploration techniques. Besides large quantities of gas already available in India from the Krishna-Godavri offshore basin, gas from shale deposits will become available all over the Gangetic plain, Gujarat, Rajasthan, NE Assam, and some coastal regions. New gas deposits are being constantly found offshore.

Gas extracted from shale deposits is definitely likely to benefit India. India has a strange restriction at present against harnessing non-conventional sources. Firms are allowed to produce conventional oil and gas, but they have to keep their hands off coal gas, methane and shale gas even if they discover them in their exploration blocks. Since gas can be a cheaper and cleaner alternative to oil, a policy change is urgently required to allow exploitation of shale gas and methane without delay.

Shale gas appears to have a bright future. Firstly, as stated above, shale gas is available at several places in India. Its extraction using the "fracking" technique is being perfected in USA. At least two Indian firms are learning those techniques for application in India. So it should be a matter of time before success can be claimed in transfer of technology to India. As mentioned earlier, the use of shale gas is a distinct possibility in the near future.

2.8.2 Wave Energy

Another important form of renewable energy that appears to have a good future in India is wave energy. Not much research work, however, appears to have been done as yet in India with this system although India has a long coast line.

The wave potential is said to be around 1,000 times more than the wind potential. Waves are a stable and predictable source of power, taking into consideration diurnal and seasonal variations. Wave energy results in a substantial quantum of power generation since the energy generated varies according to the square of wave height (as compared to wind power which varies according to cube of air speed. In addition, water is some 850 times denser than air. Moreover, infrastructure requirements are said to be much less, since wave energy needs only 1/200 of the land area of wind energy. Generating wave energy entails lower infrastructure costs.

According to an UNFCCC update (EB 51) four different technologies are under development in USA and elsewhere. Among the mature ones, is the "Terminator" installed near shore in California and is able to produce electricity at about 10 cents/kWh. The principle is illustrated below in Fig. 2.3. Portugal and Philippines are also reported to be working on these devices.

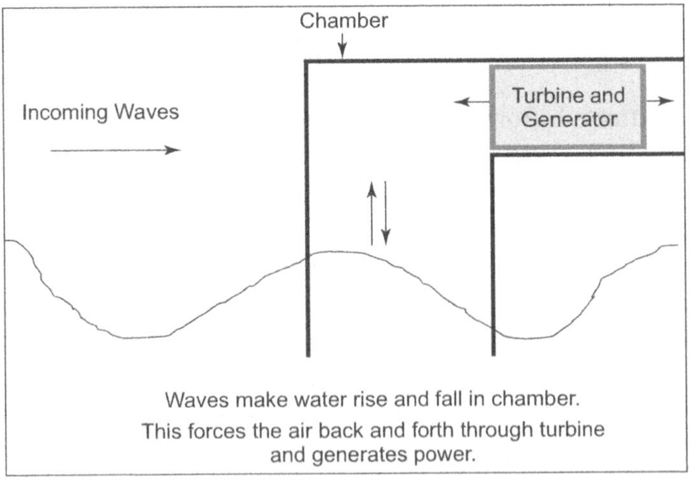

Fig. 2.3 *A turbine device to generate electric power from wave energy*

No project on wave energy harnessing yet appears to have been set up in India in spite of its long coastline and the presence of important ports and cities on the coasts.

2.8.3 Geo-thermal Energy

Geo-thermal energy is available in certain parts of the world, e.g., Iceland and around southern Russia and can be used in the form of hot or lukewarm water in homes or public baths, welcome in a cold country. It is reported that 25% of Hawaiian electricity is geo-thermal. In Iceland, nearly 80% of all homes are heated by geothermal energy.

Naturally available hot water from a groundwater source can be, for example, piped and supplied city-wide through a network of pipes laid in tunnels, built to facilitate their maintenance. The Chinese city of 5 million people, Xiang Yang, is hoping to set up one of the world's largest urban geothermal heating systems.

India is supposed to have a large scope for setting up geothermal energy projects, but no progress appears to have yet taken place in this direction to find out if any scope really exists. Mapping seems to be first needed.

Satellite-based Solar Powered Energy

The reader must also be reminded of the topic of Satellite-based solar powered systems mentioned earlier which needs some R & D work to bring to fruition. It would avoid India's land problem. Moreover, from a satellite, solar energy can be tapped 24 hours a day. This would greatly improve its efficiency.

Our Preferences

This chapter tells us that technology for many alternative sources of energy exists, but their development is slowed down at present by various factors such as costs, land requirement, etc. Already, there are some niche applications where costs can be competitive, or subsidies are possible to secure. The chapter also tells us in which direction our research programme has to focus.

A question that one may ask at this stage is: which system of all those described above might catch the fancy of the Indian public? The answer would perhaps be one that meets the following commonsense criteria at all times would be likely to become the most favoured system and become the most sustainable too:

- ➢ It should be easy to install, safe to use with existing equipment and sustainable at all times (no need to import).
- ➢ It should give the best financially attractive solution to the energy problem.
- ➢ It has to be reliable (political security is important, especially at time of war).

It is clear that gas has a great future (perhaps shale gas). Wind, solar, hydro and wave energy systems are also most likely to become the favourites in the future, depending on their relative pricing policies.

For traffic, biofuels are ready and waiting to be used with conventional vehicles provided some initiative is taken to make them available at all petrol pumps in India. Electric vehicles are likely to replace conventional ones in the more distant future. As we said earlier, in both transport and power generation, the future belongs to alternative sources of energy, not fossil fuels.

REFERENCES

Arceivala, S J, (2010), *Global Warming and Climate Change*, Indian Environmental Association.

Arceivala, S J, (2011), *Green Energy*, Forum of Free Enterprise, Mumbai.

EXERCISE QUESTIONS

1. Do you think that wind, solar, and hybrid are the best alternative sources of energy to consider today? If so, why?
2. Bio-fuels have a future only for a short while as additives to petrol or diesel as long as we continue to use our existing type vehicles. Comment.
3. What do you think would be the advantages and impediments, if any, to a change over to electric cars in near future?
4. What thoughts do you have to express on land acquisition in India? Give reasons.
5. What is an "integrated" solution? How does it differ from an "optimized" solution? Explain with examples.

PROJECT/ASSIGNMENT TOPICS

1. To add more water supply to Mumbai, Chennai and Kolkata it is said that desalination will have to be considered in near future. If so, which source of energy seems most promising to you to use for the purpose? And why?
2. Discuss the potential for reducing the carbon footprint of your city? Identify the steps to be taken.
3. Briefly discuss each of the following options for implementation, marking those which you think are more feasible to adopt, and those which might possibly have undesirable side-effects.

   Soft Options

   - Monitor automobile and industrial emissions.
   - Appoint more monitoring staff.
   - Levy more stiff fines.
   - Legislate to place a cap on electrical consumption.
   - Hold seminars and conferences.

   Hard Options

   - Provide better quality of fuel for automobiles.
   - Provide bio-fuels for admixture.
   - Manufacture better, more efficient car engines and car batteries.
   - Provide flyovers to ease traffic congestion.
   - Improve town-planning and road engineering.
   - Provide adequate mass transportation facilities.
   - Assist industries through incentives to reduce carbon emissions.
   - Implement a program to train the trainers.
   - Accelerate the country's research program.

3 CHAPTER

Rainwater Harvesting in Urban and Rural Areas

3.1 Rainwater Harvesting in Urban Areas, and its Hazards

3.2 Rainwater Harvesting in Rural Areas, and its Advantages

3.3 Some Common Aspects to Urban and Rural Supplies

3.4 Summary of Recommendations

Rainwater harvesting can become a small but crucial way of adding to our water resources on a self-help basis, but beware, it has health hazards of its own.

3.1 RAINWATER HARVESTING IN URBAN AREAS, AND ITS HAZARDS

3.1.1 Introduction

Rainwater harvesting is a self-sustained and privatized form of water supply where each one is his own master and quality monitor. It works on a "do-it-yourself" basis for the owner and "leave-us-alone" basis for the local body. Therefore, responsibility for quantity and quality assurance is attached to the owner for this activity. How has rainwater harvesting to be done so that one's minimum water requirement is met with reasonable convenience and, more importantly, one's health is protected?

The water resources of the country are slowly but surely running out and urgent steps need to be taken to conserve them, and supplement them in whatever way we can. Easily available river waters have traditionally been used first to meet domestic and industrial demand. Wherever groundwater has been available, the people have had similar recourse to its use. Even this has had its limitations.

In Delhi, the groundwater level is reported to have fallen by as much as 32 feet in the last decade. In Saurashtra, it is reported to have fallen by as much as 500 feet. Over a 1000 ponds are reported to have vanished in the Meerut district. In Mumbai and some other metropolitan areas, it has become common practice to supplement dwindling municipal supplies by private water supply tankers. The unfortunate result is that we supply water of dubious quality to the people while using precious imported diesel oil to distribute the same. Rainwater harvesting is a better way to add to our water resources.

Harvesting and the related methods to be used have to be appropriate depending on the rainfall, the terrain (geology) in which one is located, the groundwater condition, and the density of the built-up area and, in present day, its pollution potential.

Experience in this subject is not lacking. From the latest excavations at Dholavira in Cutch, we find a good water management system existed even in the Harappan period. Several old rainwater harvesting structures have been found in various parts of India such as the water scarce regions of Rajasthan. In Tamil Nadu, the ancient people stored rainwater in ponds, separately for drinking and other domestic uses. Even today in Tamil Nadu, industries are forbidden to discharge their effluents into the open but encouraged to use modern membrane technologies to ensure as much reuse as possible in their own factories to reduce fresh water intake.

Once upon a time a developing city like Mumbai (then called Bombay) which receives good rainfall, had more than 8000 wells. Most were, however, polluted and, thus, ordered to be shut down when good quality piped water supplies became available towards the end of the 19th century. Now, when piped water supply is becoming inadequate, we are going in the *reverse* direction, asking that old wells be opened up or new ones sunk to supplement dwindling municipal supplies! This may be required from the standpoint of additional water resources for the city but we may thereby be exposing public health to undue hazards of dual quality water supply. Public health protection is after all a mandate to be followed by all local bodies. Hence, we are torn between quantity and quality of supply. We will discuss this further later.

3.1.2 Some Geological Aspects

Maharashtra

Geologically, the rocks in Mumbai and many parts of Maharashtra are primarily of the igneous type (basalt, Deccan trap, etc.). The igneous rocks have often undergone weathering action in the upper layers leading to softer murrum formation capable of holding water whereas the lower portions have remained relatively impervious. Also, in case of coastal areas like Mumbai, fresh water lenses have tended to float on heavier brackish waters below.

Cities like Pune, also have igneous rocks and the generally rocky terrains have large tracks of black cotton soils as well as fissures and crevices filled with softer soils. Additionally, Pune being on the downwind side of the Western Ghats, has much lesser rainfall than Mumbai and thus needs larger roof areas to catch enough rainfall (see later).

Matheran, Mahableshwar and all the hilly areas of the region in-between are generally characterized by heavy rainfall in the monsoon months. The result is that rainwater quickly runs down the slopes into the valleys below, leaving the habitation on the hill tops dry. The hill tops thus become perfect candidates for rainwater harvesting. Constructing dams and pumping up the water from the river valleys below to the top can be quite expensive. Thus, rainwater harvesting directly where it falls (on the hill tops) can be a very helpful and, in most cases, a very economical alternative to follow for hilly areas.

The climate, the geology, the basin morphology and vegetation in the catchment area are important variables that control the geo-hydrology of the plateau. The mass of hills (Matheran, Mahableshwar, etc.) is composed of basaltic trap which is impervious. Owing to weathering action, this mass is overlaid with a crust of soft "laterite" rock which is relatively porous and can hold water. It is also often reddish brown in colour owing to the presence of iron leached out from the rock. The water held in the laterite oozes out when it comes in contact with the basalt below, resulting in a lot of springs on the edge of the plateau. These springs are an important water resource for the plateau and the surrounding regions.

The land surface is generally irregular giving rise to various water features such as surface streams, basins, "dharas", waterfalls and springs. A high level of soil moisture results in the evergreen nature of the vegetation there. Dug wells and drawing of ground water will result in reduction of soil moisture which in turn will increase soil erosion unless replenished. Vegetation is sparse on the plateau tops.

Gujarat

In Gujarat, and many other parts of India, the characteristics are totally different from those of igneous rocks. The soils in Gujarat are often of the sedimentary type with alluvial soils predominating.

Gujarat soils are generally more water-laden and in several cases the harvested rainwater can be transferred to the soil more readily for holding until later use. Sea water intrusion can also happen more readily in coastal areas leading to brackish waters as more and more sweet (fresh) ground water is withdrawn. Pollution can also spread more easily over wider areas once introduced into the ground water.

3.1.3 Some Basic Design Aspects

What follows is applicable to many parts of the country. Some details are given below on how rainwater can be harvested, supplemented with groundwater, treated if necessary, and used without danger to health.

Simple water conservation measures and even reuse of wastewater after treatment also become important complimentary measures to add to the water resources of such areas and have, therefore, been included here.

Water Requirement

Generally, the water storage requirement depends on the number of users and the use to which the stored water is to be put. The approximate daily requirement of water in liters per person for different uses is indicated below in Table 3.1 on the assumption that where rainwater harvesting is necessary one will tend to be economical in use of water.

Table 3.1 *Water requirements for different purposes in individual houses using harvested rainwater*

	Use of Water	Estimated Daily per Capita Requirement (L/Cap/day)
1.	Only drinking	2
2.	Drinking, cooking and minor uses	10
3.	Bathrooms, washing, cleaning, etc.	40–50
4.	Toilet flushing	~ 40
5.	Total for household	~ 90
6.	Total for household, with reuse	~ 40–50

Note: Gardening requirements can be estimated on the basis of 7–10 $L/m^2/day$.

We see from Table 3.1 that the water requirement for drinking alone is only about 2 L/person/day on an average. This requirement increases to about 10 L/person/day if cooking and minor household uses are to be covered besides drinking. Thus, minimum water storage required in RWH areas varies from 2–10 L/person/day.

Additionally, however, 40–50 L/person/day are required to be stored where harvested water is to be used for bathing, clothes washing and cleaning. Where the toilets are cistern-flushed, yet another 40 L/person/day would need to be stored. Thus, if all uses have to be met (except gardening) the full quantity of ~ 90 L/person/day has to be stored. This requirement can be reduced to half the volume if the so-called "gray" waters from bath, etc., are reused (after some treatment) for toilet flushing or the toilets are of the hand-flush type requiring minimum water (only about 8–10 L/person/day).

For hotels, guest houses, and even a few individual houses that can afford it, much larger quantities than those given in the table above, may be stored provided water is available and storage is affordable. For municipal water supply systems in urban areas much higher norms are used as discussed in Chapter 6.

3.1.4 Other Parameters in Design

The other parameters in design are interrelated and are given below:

- ➢ The number of users
- ➢ The annual rainfall and its available fraction
- ➢ The roof collection area
- ➢ The number of dry days for which storage is required
- ➢ Cost of storage tank construction

The storage required

$$= \text{No. of users} \times \text{daily consumption rate} \times \text{No. of dry days}$$

If daily consumption rate is 90 L /cap-day (full service), the number of users are, say 10 and the number of dry days are 300,

The storage required is

$$= 10 \text{ users} \times 90 \text{ L/person/day} \times 300 \text{ d} = 270{,}000 \text{ L per year}$$

Number of Dry Days

It would be safe to assume in the western region that the number of dry days in a year will be 300 for which storage will be required.

Rainfall and Roof Area

Rainwater available is generally assumed to be 80% of the annual rainfall in the region and is multiplied by the roof area to give the available harvested volume.

In the Mumbai region, for example, the rainfall is assumed to be 2000 mm annually while in some of its hilly regions it may be more. In the Pune region it is only 600 mm annually.

Hence, to harvest the above calculated water requirement of 270,000 L, the roof area required in Mumbai region = 270,000/0.8 × 2000 = 170 sq m approx.

To harvest the same volume in Pune region, the roof area required is = 563 sq m. If the roof area of the house is lesser than required, one has to make do with lesser volume of water.

Cost of Storage Tank

Cost of storage tank is yet another parameter to be considered. The cost may vary from ₹ 2.50 to ₹ 3.50 per L capacity of tank. For a 270,000 L storage tank the cost would be nearly ₹ 10 lakh. Is this cost affordable for a family of 10 persons?

The affordability of storage tank cost often determines the capacity stored. One could even store a larger volume if money, roof area and space for storage are available.

Rainwater Harvesting Arrangement

In general, rainwater harvesting (RWH) is usually done in either one of the following two different ways:

1. *To serve a household, directly with harvested rainwater with no groundwater supplementation*

   This may have to be done either in case of a rocky hill-top terrain with no possibility of groundwater supplementation, or in other types of terrain where for reasons of quality groundwater supplementation is not desired.
2. *To serve one or more households with harvested rainwater along with groundwater supplementation*

   The harvested rainwater (if collected hygienically) generally needs no treatment and can be supplied directly for potable purposes. If groundwater is used for supplementation, however, it may be contaminated or polluted and may need some treatment. Thus, wherever groundwater and the harvested rainwater are to be mixed, the resulting quality is more akin to the groundwater quality and may need treatment before use or may have to be restricted to non-potable uses only.

The treatment methods applicable are mentioned later below.

Rainwater Harvesting with no Groundwater Supplementation

Where no groundwater supplementation is done, rainwater is generally stored in a masonry tank of a size big enough for the full quantity required during the whole dry season.

Evidently, the capacity of the storage tank has to be based on as high a daily consumption rate as may be affordable and feasible at site. Preferably, the quantity should be adequate for meeting the full requirement of water for all purposes such as drinking, cooking, bathing, washing, flushing, etc., i.e., 90 L/person/day all directly from the harvested rainwater (see Table 3.1).

If the "gray" waters from bathing, washing, etc., are treated and reused in toilet flushing, the daily per capita consumption can reduce from 90 to 40 or 45 L/person/day. This would reduce the rainwater storage tank capacity by a substantial 50% or so. However, this saving would be partly offset by the additional cost (and bother) of operating the reuse system.

A typical reuse arrangement is shown later in Fig. 3.1.

As mentioned earlier, the water collected from the roof needs no special treatment and only the stored water needs to be protected.

Rainwater is assumed to be clean and drinkable. However, as industrialization progresses and air pollution control is not adequate, rainwater may carry a few particles of substances such as the heavy metals and gases emitted from industrial sources nearby. Normally, these pollutants are in relatively insignificant concentrations and often neglected.

Out of past experience, rainwater collected on rooftops is provided no special treatment except to make sure that a bye-pass connection is provided in the downtake pipe so that the roof or terrace surface can be flushed clean of fallen leaves and dust by the initial rains before the water is admitted to the storage tank (see Fig. 3.1).

The downtake pipes from the roof (often provided according to the thumb rule of 1 per every 4 sq. meters of roof area) are led to a simple sand and gravel filter for coarse filtration before the water goes to the storage tank so as to exclude any wind-blown leaves and dirt which may have come on the roof. Limestone is often used to keep the stored water on the alkaline side for improving its keeping quality or else bleaching powder is used occasionally to disinfect the water. The tank is fully enclosed and all apertures are screened to prevent access to mosquitoes, rodents, etc.

Rainwater Harvesting along with Groundwater Supplementation

In this case it is generally easier to meet the full service requirement of 90 L/cap-day as only a part of the water requirement has to be harvested from the roof and the other part is obtained from the groundwater available in the area.

Only the rainwater harvested from the roof is stored directly in a masonry tank (i.e., generally the drinking-cooking requirement of 2–10 L/person/day) and the excess quantity is transferred to the underground water-table through an open well or bore-well or a "recharge well" for later use provided the soil is suitable for holding the water. This reduces the required storage tank volume and, therefore, the overall cost.

For example, even though the rainwater available at a site during the wet season may be more than adequate to meet the full requirement of 90 L/cap-day for all household purposes, one may build the storage tank for, say, only 10 L/cap-day and let all the surplus rainwater overflow from the storage tank to a well in the compound to raise the general aquifer level (see Fig. 3.1).

> The same recharge well can then be used later to pump back the groundwater to the house as and when needed, but it should be remembered the groundwater may be polluted or need some treatment, or may have to be confined to non-potable uses such as toilet flushing, gardening, etc., through a separate piping network. (Also see Fig. 3.1)

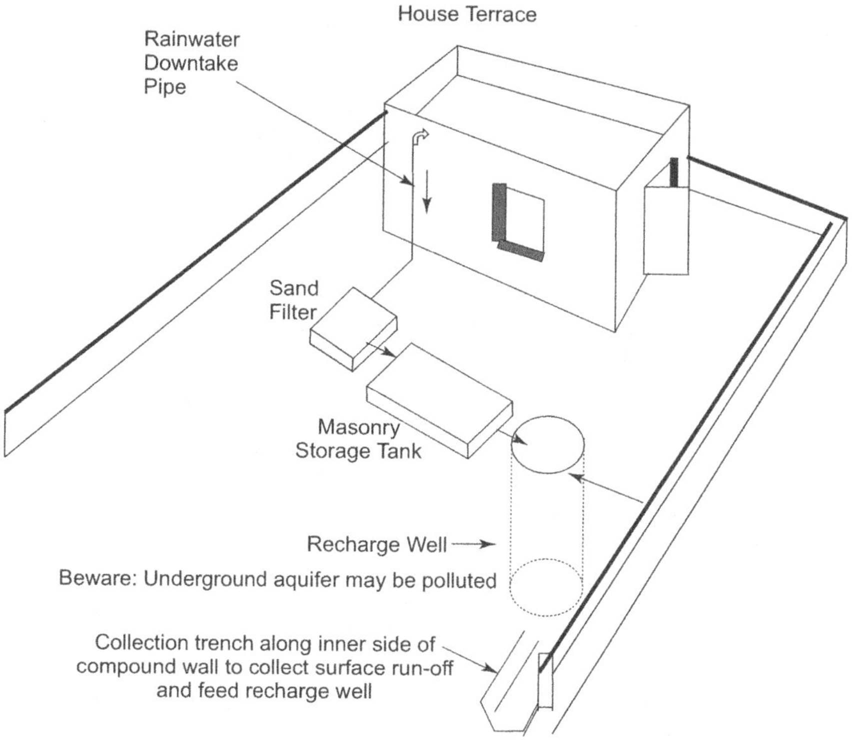

Fig. 3.1 *Rainwater harvesting along with groundwater supplementation for individual households*

Additionally, at some further cost, the rainwater falling on the whole surface of the plot of land can also be drained into the same well or into a separate recharge well (if contours require it) for percolation into the ground by providing a gravel-filled collection channel or trench around the plot as shown in Fig. 3.1. This would bring in even more water into the ground but, as stated earlier, the water quality may not always be fit for drinking without further treatment.

The advantage of being able to recharge ground water and then use it later is that one could meet the requirements for washing, bathing, laundry and flushing of toilets from groundwater although the storage tank is of a small size. Gardening needs can also be met if sufficient groundwater is available. All this is possible only if the soil is favorable for holding water and one is prepared to treat the groundwater as may be necessary or use it for non-potable purposes.

This is what is presently being recommended by the Brihanmumbai Municipal Corporation to be done in Mumbai and some other urban areas though the reader should be aware of its limitations. As mentioned below, groundwater recharge would be useless in certain cases in Mumbai. Before such a scheme is developed it is desirable to be sure of at least two things:

1. Make sure the roof-water percolated underground does not drain off quickly to some other catchment or site and leave the proposed "beneficiary" high and dry. This could happen, especially in a hilly region with unfavourable geological formations.
2. Also make sure the existing groundwater is of acceptable quality for human consumption as otherwise one would be adding "good" water to "bad" water and then spending a large amount of money on its treatment or on installing a separate network for distributing non-potable water.

Mumbai Groundwater Quality

As is well known, Mumbai was originally a group of 7 islands with the sea all round. In 1967, a VJTI environmental engineering student surveyed 76 existing wells in all of Greater Mumbai to determine their quality, as a part of his thesis work. His findings were very interesting:

- Some wells showed salt water intrusion (brackishness).
- Some wells had high hardness and presence of chemicals (industrial pollution).
- Some wells showed high chlorides (sewage pollution).
- Some wells had a relatively shallow layer of sweet water floating on brackish sea water.
- The Northern suburbs had fractured rocks permitting excessive leakage.

Harvested rainwater from storage tank can be used directly for potable purposes without treatment. The overflow from storage tank and the surface runoff collected by trenches along the compound wall, can be used to recharge ground water through a well. At other times the same well can be used to pump up ground water for supply to the house for non-potable purposes through a separate piping network.

> Groundwater recharge by harvested rainwater cannot be done everywhere in a city like Greater Mumbai. In some cases, the rainwater would just mingle with the sea water. In some other cases it would become polluted with sewage or industrial wastes and need heavy pre-treatment before supply. In some cases it would just be lost underground. Thus, a geologist and some lab testing are essential to ensure groundwater recharge is done only where it is truly feasible.

3.2 RAINWATER HARVESTING IN RURAL AREAS, AND ITS ADVANTAGES

Rainwater harvesting for augmenting community-level water supplies is often necessary in rural areas and small towns that depend solely on ground waters. The whole community then benefits. The wells dry up within a few months after the monsoon is over and the people have to make do with very little water or traverse long distances to fetch water for their daily chores. In such cases, rainwater harvesting is useful to undertake for augmenting community-level water supplies by improving the moisture regime over a wider area and raising the general ground water level in the area so as to benefit not only a cluster of houses or a farm, but perhaps a whole village or community.

Rainwater harvesting for such a purpose is usually a relatively large-scale operation and therefore, it generally involves the local body and perhaps an NGO or two and becomes a fit case of "public-private participation". This imposes a socio-economic dimension on the problem and succeeds well where there is adequate leadership.

> **The Power of Water**
>
> When the wells can function for longer periods, the farmers can harvest two crops instead of one, and their entire economic condition changes. The face of India can also thereby be changed. Such is the power of water when harvested in rural areas!

3.2.1 Techniques Available for Augmenting Rural Supplies

A few of the typical techniques used to harvest rainwater in villages and rural areas are listed below and shown diagrammatically in Figs 3.2–3.8.

It is important to know the geology and topography of the area concerned. Generalizations cannot be made since some of them may be feasible, and some not, in certain regions depending on the geology and soil conditions. Hence, the inclusion of a hydro-geologist is invariably recommended. The inclusion of "water-diviners" and local experts is often useful but only as "resource" persons in addition to the hydro-geologist. Some of the available methods are listed below:

1. Percolation tanks
2. Contour leveling (terracing)
3. Ridge and Furrow construction
4. Provision of Kundis (wells) and Tankas (tanks)
5. Provision of Jaldhars (bunds) and Chaukas (pits)
6. Step-wells or Bavdis
7. Provision of check dams and creation of ponds
8. Construction of surface and sub-surface dykes
9. Use of other methods which hold allied advantages with RWH

The above methods are briefly described below and shown in Figs 3.2–3.8.

Percolation Tanks

A percolation tank is shown in Fig. 3.2 is, as the name suggests, a large, shallow basin excavated at the apex of a hill to allow rainwater to percolate into the ground. The tank may be partly filled with boulders and/or gravel to let the rainwater percolate through the interstices into the weathered rock below and reappear later down as a "spring" which can be captured and used.

Contour Leveling or Terracing

Contour leveling or terracing is done (like in a tea garden) to slow down the speed of flowing runoff and encourage it to enter the soil below (see Fig. 3.3). It also reduces scope for erosion of soil. To minimize the earthwork, the "cut" portion and the "fill" portions are more or less balanced. This is commonly done on hill slopes.

Ridge-and-Furrow Type Construction

Instead of terracing one could have a "ridge-and-furrow" type arrangement on the slope where the "ridges" would slow down the flowing water and encourage diversion and percolation through the trenches or "furrows". The natural slope of the hillside is thus retained (see Fig. 3.4). This arrangement also tends to reduce erosion.

Collection Wells or "Kundis"

If the topography is suitable, a "kundi" or a shallow collection well can be built in within a depression in a given area provided the soil underneath is capable of holding water (see Fig. 3.5). The openings or inlets for water to go into the well are guarded by a wire mesh to prevent the entry of floating debris, birds and reptiles. The slope towards the well would increase the yield from the catchment area. To increase the yield further, the surface can be treated by using clay to make it impervious. After all the rainfall in parts of Rajasthan may be of the order of 25 mm on at least 4–6 days in a year.

In China, polythene sheets are used to cover wide areas and increase the surface flow contributing to the collection well. In India too, polymer science has been used to reduce permeability of sandy soils and thus increase output.

A "tanka" or small tank is commonly seen in northern Gujarat and Kutch areas for similarly collecting rainwater and storing it for use as drinking water during the dry months. The tankas are often lined with fine lime to protect and preserve the water.

"Chaukas" (Infiltration Pits) along with "Jaldhars" (Shallow Bunds)

"Chaukas" and "Jaldhars" are Rajasthani terms used in rainwater harvesting. Shallow dykes, bunds or Jaldhars are constructed to hold water at the lower end of a sloping plot of land while several chaukas (shallow pits 1 m × 1.5 m × 1.5 m) are constructed within the area to encourage percolation below ground. The collected water soaks into the soil to increase soil moisture (see Fig. 3.6). The method is commonly used even where the slopes are more or less flat.

Shrubs are planted on the "Jaldhars" to give them additional support to withstand rain. When it rains, the "chaukas" are filled one by one along the slight slope and finally enter a drain providing water for the villagers and recharging groundwater – an ecologically perfect solution.

Step-wells or "Bavdis"

Step-wells or "bavdis" as they are called have been used where the ground water is available at some depth. Such wells are now avoided owing to the potential for spread of guinea worms from a sufferer stepping into the well water and spreading the disease to others.

Provision of Check Dams and Creation of Ponds or Network of Ponds

Small masonry check dams, anicuts or weirs of locally available material can be built across small surface streams to hold back the flowing rainwater and encourage downward percolation into the aquifer (see Fig. 3.7(a)). Alternatively the runoff may be held in small ponds especially in mountainous regions (see Fig. 3.7(b)). The weir foundations must be stable. If topography permits, a set of several ponds may be created and the ponds linked together into a network where possible.

Such interventions carried out above ground to intercept and slow down flowing streams has led to "miracles" in the form of increased availability of water for drinking and agriculture to the local people.

A variation of the above are the tanks called "Kere" in some parts of Karnataka used for storing waters deflected by check dams across streams to serve the irrigation fields on either side. The tanks are often built in series so that the seepage from the higher tanks flows into the lower ones and no wastage occurs.

Surface and Sub-surface Dams or Dykes

Figure 3.8 shows an example of a small dam constructed below ground level to raise the level of sub-surface groundwater flow. Investigations are first necessary to ensure its feasibility. A hand-pump can be used to lift up the water to ground level where levels permit.

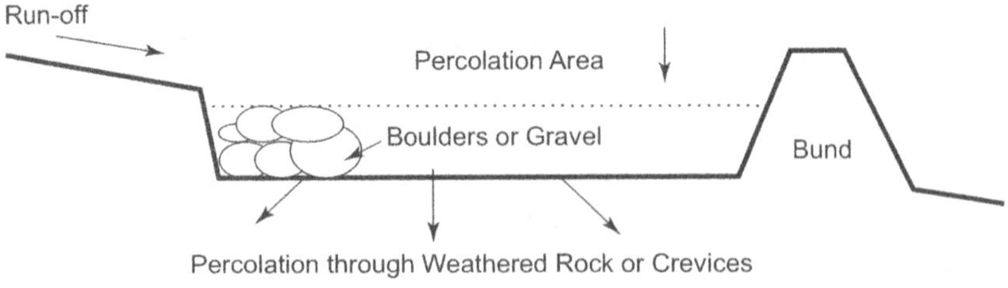

Fig. 3.2 *A percolation tank*

A shallow, percolation tank filled with boulders and/or gravel may be perched on selected high ground to feed rainwater run-off to weathered rock or crevices below so that the water emerges somewhere further below as a spring.

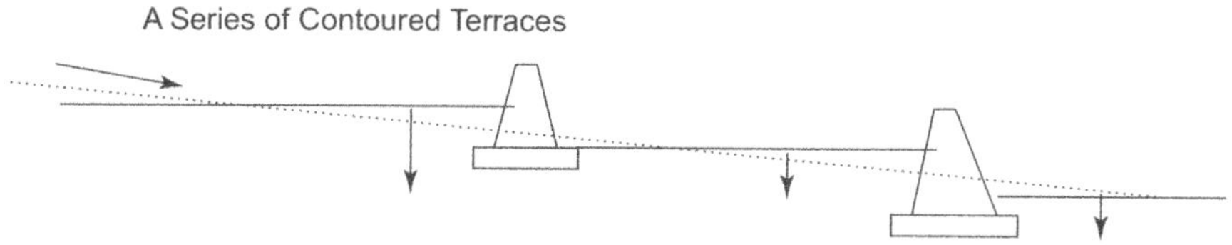

Fig. 3.3 *Contoured terraces*

A sloping hillside may be converted into contoured terraces (as in a tea garden) to slow down the velocity of flowing water and encourage percolation into the hill. It would also prevent soil erosion. The cut and fill portions are more or less balanced to minimize earthwork.

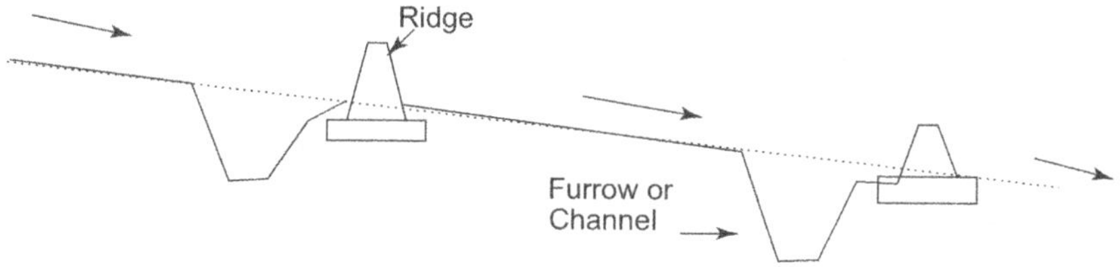

Fig. 3.4 *Ridge and furrow type construction*

A ridge and furrow type construction may be adopted in hilly or sloping areas to slow down or arrest the flowing water at the ridges and convey it via trenches or furrows to encourage percolation into the ground. The natural slope of the hillside is retained.

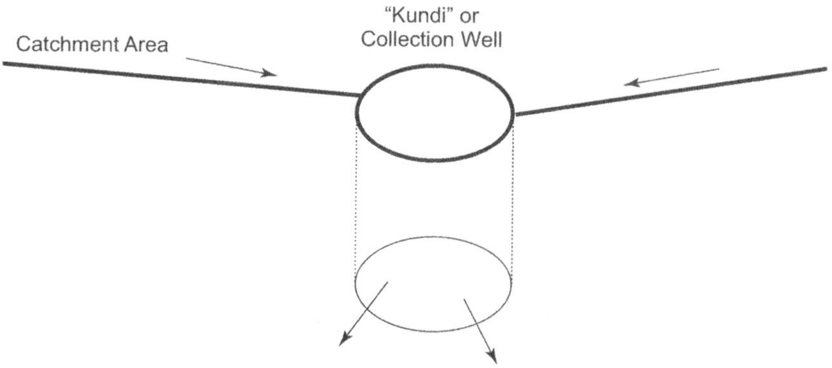

Fig. 3.5 *Provision of "Kundis"*

A "kundi" or collection well can be built at a low point in a depression to help drain rainwater into it and thus feed the aquifer below. To increase the drainage from catchment area, the surface may be treated by clay to make it impervious. In China, polythene sheets are sometimes used instead of clay.

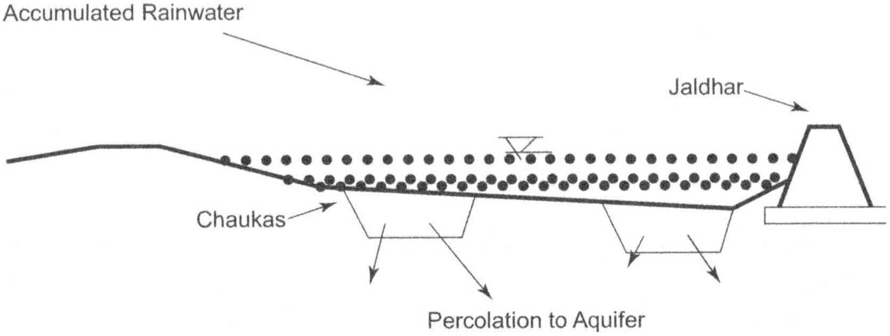

Fig. 3.6 *Small bunds or "Jaldhars" to hold back the flowing rainwater and "Chaukas" or infiltration pits for percolation of the water to the aquifer*

"Chaukas" and "Jaldhars" are Rajasthani terms used in rainwater harvesting. Shallow dykes, bunds or Jaldhars are constructed to hold water at the lower end of a slightly sloping plot of land while several chaukas (shallow pits 1 m × 1.5 m × 1.5 m) are constructed to encourage percolation below ground. The collected water soaks into the silty soil below to increase the general soil moisture content.

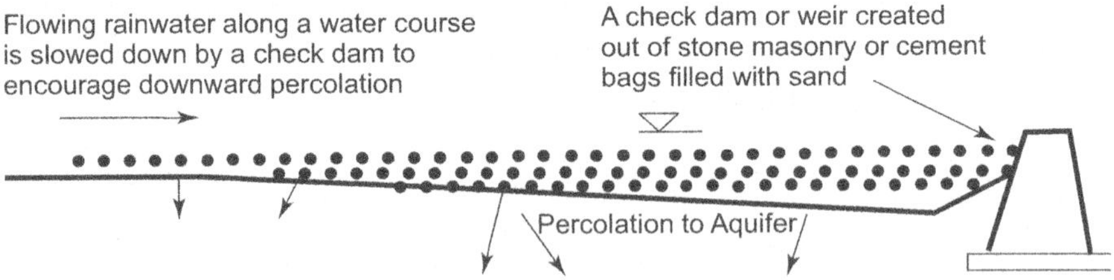

Fig. 3.7 (a) *Provision of a check dam in Maharashtra to encourage downward percolation of rainwater*

Where topography permits and the soil is suitable, a masonry weir may be built to create a pond or a network of interconnected ponds or even a lake to Hold rainwater. The weir foundation must be stable.

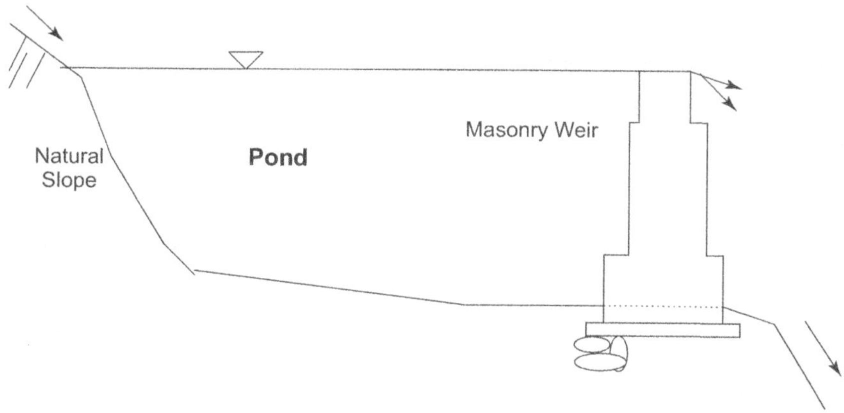

Fig. 3.7 (b) *Provision of a check dam in a mountainous terrain-creating a holding pond*

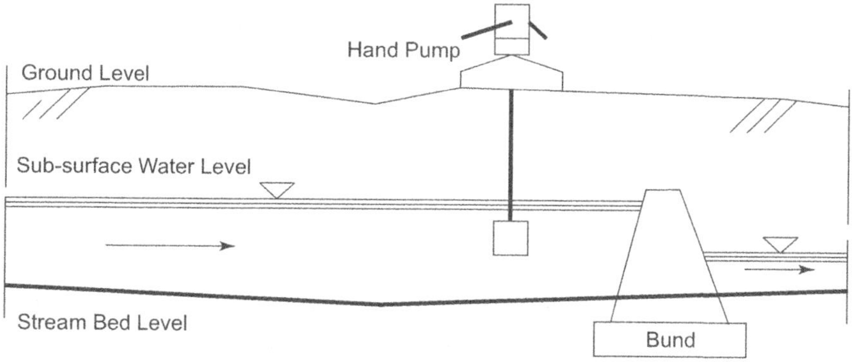

Fig. 3.8 *A sub-surface bund or check dam*

A sub-surface bund may be constructed to slow down groundwater flow and raise its level so as to become accessible by a hand pump. Similarly, a bund may be constructed across a surface stream to slow down the flowing water and raise its level to promote ground water recharge (as shown in Fig. 3.7(a)).

3.2.2 Approaches Involved in Augmenting Rural Supplies

From the various methods listed above it will be seen that, essentially, there are two approaches, one direct and the other indirect, in augmenting rural water supplies through rainwater harvesting.

The Direct Approach

The basic principle underlying such rainwater harvesting systems is to cut down the speed of flowing runoff water and give it more time to settle and percolate underground.

Forests and trees have long been known to be useful in reducing the speed of rainwater run-off and consequently increasing percolation and even reducing erosion. Forests and trees are by themselves not always enough in urban or hilly regions and need to be supplemented by the various techniques noted above to give water more time to percolate.

> It can be said, in a lighter vein, that
>
> "If water is running, make it walk. If is walking, make it stand. If it is standing, make it sit. If it is sitting, make it sleep. By using various "tricks", give water all the time to percolate underground!"

The Indirect Approach

As the name suggests, this is an indirect approach where something is done in the hope that it will indirectly promote rainwater harvesting.

People's ingenuity has enabled them to devise various ways of handling the rainwater harvesting problem. One such device is to explore use of other methods which hold allied advantages with rainwater harvesting. For example, China has built good surfaced roads in villages as they would not only provide communication for transport of the agricultural produce, but the hard surface of the roads would also enable some rainwater harvesting to be done and its conveyance via roadside channels to use in agriculture.

In India, Swaminathan S Anklesaria Aiyar strongly believes that the way out of rural poverty is through capital-intensive rural development schemes rather than labour-intensive ones. According to him, many studies have shown that with the provision of all-weather roads (which need capital-intensive construction) and water and other infrastructure, development occurs rapidly because businesses can spring up everywhere and provide new jobs. Greater attention needs to be paid to this aspect than done hitherto in our country.

> S Anklesaria Aiyar says:
>
> "The trick lies in exploring the *dual or co-lateral advantages* between water harvesting and other economic activities so that one automatically promotes the other".

3.3 SOME COMMON ASPECTS TO URBAN AND RURAL SUPPLIES

This section hopefully shows that water quality makes a great difference to the value of a project and distinguishes it from an amateurish attempt often made to meet water shortage without any thought for water quality and health protection. Many talk of rainwater harvesting; few talk of water quality. In a holistic approach, no rainwater harvesting scheme can be considered without first considering the quality of groundwater.

As discussed earlier, groundwater used for supplementation always has to be adequate for human consumption whether it is in an urban or rural area. Generally, the quality of existing groundwater gives a good idea of what will happen later. Rainwater harvesting improves in a limited way the quality of groundwater because of the relatively little dilution it provides to the existing groundwater when added to it. Thus, rainwater harvesting as such is never the primary cause of poor quality. It is always the quality of the groundwater used for supplementation that matters.

3.3.1 Some Thoughts on Analysis and Treatment

A bio-chemical analysis of the existing groundwater quality is always desirable to carry out, both in urban and rural areas, since the groundwater may be unduly high in any one or more of the following type of constituents:

1. Hardness
2. Iron and manganese
3. Fluorides
4. Nitrates
5. Arsenic
6. Upstream pollution from sewage or industrial wastes

Polluted groundwaters may require pre-treatment to meet drinking water standards which may sometimes work out to be more expensive and bothersome than building a larger storage tank to store all the rainwater required over the dry season. Sampling and analysis of the groundwater would reveal its quality. The decision to use groundwater supplementation, then depends on local circumstances, costs to be incurred to meet drinking water quality standards and personal preferences.

Fortunately, all the pollutants listed above are not found at the same time. For example, hardness due to calcium carbonate or calcium sulfate in the soil is a more commonly occurring problem in many parts of India, including Maharashtra and Gujarat. Iron and manganese are also prevalent in many parts especially including eastern and north-eastern India. Fluorides are found mainly in Andhra Pradesh, Tamil Nadu and some parts of North India while arsenic is mainly the problem of West Bengal.

Treatment

Details of treatment methods are not given here as they are available in standard text-books and manuals.

> It is sufficient to say that in choosing between the different treatment methods, preference has to be generally given to those methods which are operationally easier to handle in a given case. Thus, methods in which chemicals have to be measured, added, stirred, etc., are often avoided and preference given to resin beds, pressure sand filters and such items through which the water to be treated is only required to be passed, preferably by gravity as electric power may not be available all the time.

Pollution from Leaking Sewers in Urban Areas

Pollution from leaking sewers in urban areas is a serious problem. It is a well known fact that in all the cities of the world (including Mumbai, Pune, Ahmedabad and other cities of India) where sewers exist, sewage often leaks from somewhere or the other. Thus, the upper few meters of soil in any city should always be suspected of pollution. Industrial wastes contained in the sewage can only make matters worse.

Wherever groundwater is proposed to be used in a city, it may either have to be treated properly to make it potable before use or used only for non-potable purposes. The latter would require a separate network of piping from the one used for potable water supply and the danger of a mix-up between the two systems cannot be ruled out. The track record of plumbers in many countries shows that a mix-up between the two is easily possible. Is health then protected? It is also not possible to fix responsibility in case an epidemic occurs. Hence, it is safest to have the non-potable supply also treated as good as drinking water though it is supplied through a separate network of pipes.

In rural and touristic areas where population densities are low, the problem of people pollution may not be so acute but, nonetheless, the dangers are present and thus a simple form of "environmental impact study" may be needed in some cases to ensure public health is protected.

Where the pre-treatment required is relatively easy to give at the individual consumer's level, (e.g., simple disinfection) it may be so done. However, if the pre-treatment required is more elaborate, it has to be preferably done at the community level before water is supplied. Treatment for removal of hardness, iron and manganese, fluorides and such other impurities (including sewage pollution) are generally done at community level before supply.

The State government of Tamil Nadu has made rainwater harvesting mandatory in the whole state since October 2003. Brihanmumbai Municipal Corporation requires all new buildings in Mumbai city over a certain size to install arrangements for rainwater harvesting. The reader is referred to earlier discussion regarding the limitations of groundwater quality in Mumbai. Besides Chennai and Mumbai, some efforts are also being made in Delhi and other Indian cities to benefit from rainwater harvesting. Local requirements must be met.

Costs

The costs of rainwater harvesting systems are very variable depending on the terrain, geology, rainfall, depth of groundwater, extent of storage, type of treatment required, pumping, and such other local features. Thus, it would be somewhat misleading to indicate a definite amount for an urban or rural scheme. One would have to cost out the project cost on the basis of its components.

For urban schemes the essential components are:

1. Rooftop collection, downtake, bye-pass, sand filter costs.
2. Storage tank, flush-out, air-vent, overflow pipe costs
3. Recharge well, groundwater pump-up
4. Groundwater treatment costs, if any
5. Internal supply piping for potable and non-potable supplies

For rural schemes the essential components are:

1. Cost of percolation tanks or contour terracing, constructing bunds, channels, dykes, surface or sub-surface check dams, weirs, etc., for rainwater harvesting depending on site conditions.
2. Cost of providing hand-pumps, dug-wells, bore-wells, etc., as necessary.

Several small schemes in Maharashtra have cost anything between ₹ 2 lakh and ₹ 10 lakh each, and some have cost much more. Some schemes have been carried out with local funds, some with NGO assistance (e.g., Rotary Clubs), and some at governmental cost. In view of the large number of variables involved, it is not desirable to indicate any firm costs without knowing the exact site conditions and other related parameters.

3.3.2 Water Conservation

For obvious reasons, water conservation has to go hand in hand with rainwater harvesting. One cannot have the latter without the former. Every conceivable water conservation measure has first to be adopted to reduce the primary demand for water before rainwater harvesting can become meaningful. There are two main players in this field: the general public, and industry. Local municipal authorities have a key role to play in spreading awareness of the ways in which water can be conserved. Some methods have already been discussed in Section 1.4 on water conservation.

Industries generally have a high degree of motivation for water conservation as their factory production depends on it. Two examples from Mumbai are often quoted: Years ago, textile mills found they could save as much as 15% of their water intake just by adopting certain water conservation measures including reuse of wastewater, without any treatment, from one process to another process where water quality was not important. With treatment, they could conserve a much higher percentage.

Another example, from recent times, is of the Asian Paints factory in Mumbai where water consumption was reported to have been reduced remarkably from 2800–1300 liters per tonne of production by adopting various water conservation measures. This incidentally saved crores of rupees annually in water charges.

Rainwater harvesting was then started from 2002 at its Bhandup plant by conveying rainwater from selected rooftops and collecting in Syntex tanks for use in the factory. Nearly 70 lakh liters were thus collected and used during the monsoon months.

Reuse of Water

Besides the above water conservation measures, treatment of wastewater may also be done for reuse so as to have additional water conservation.

Treatment of "gray" waters from baths and kitchens for reuse in gardening or irrigation of crops is often done together with the so-called "black" waters from toilets, using simple, non-mechanized "*natural*" *methods* of treatment described in Section 7.3.

Alternatively, the "gray" waters may be used directly for gardening and the "black" waters may be treated by any of the above stated methods and used for gardening or crop irrigation. While such reuse puts the treated effluent to good use, it does not reduce the basic consumption of fresh water.

In order to get the benefit of maximum conservation of fresh water, the effluent from bathrooms and kitchens ("gray" waters) can be just held in a small balancing tank, with bleaching powder to chlorinate the wastewater, before direct reuse in toilet flushing. The "black" waters from the toilets can then be treated by any of the methods given above for reuse in gardening or crop irrigation so that the overall water requirement is reduced to a considerable extent (see Fig. 3.9).

In hilly terrains and in the large number of small towns in India, whether hilly or not, "decentralized" sewage collection systems have a useful role to play. Natural methods are greatly recommended to be used with decentralized systems where flows are small and population densities are low as is generally the case in small towns and hilly regions.

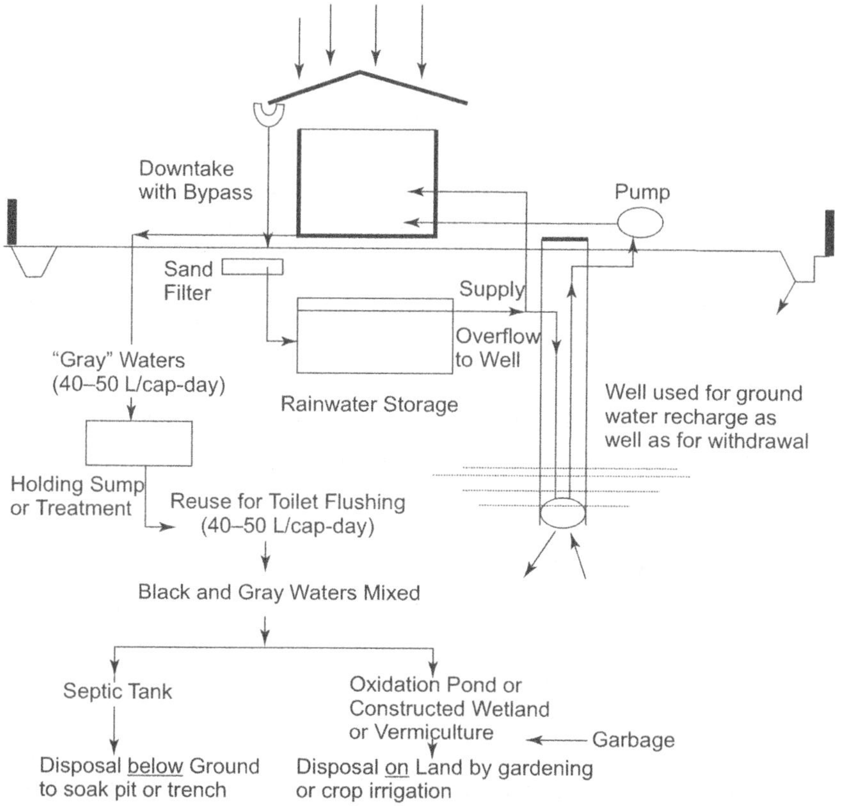

Fig. 3.9

Figure 3.9 shows how potable supply is taken from the rainwater storage tank at 2–10 L /cap-day. All non-potable uses are met from ground water which is pumped up and supplied through a separate network or treated, if necessary, before use as potable supply. The diagram also indicates how wastewater can be reused and disposed of where a public sewerage system does not exist (see text). Garbage from kitchen can be composted by vermiculture.

Moreover, it makes good sense to keep collection sewer lines short in length in a hilly topography.

Final Disposal of Treated Wastewaters

The common way in which treated wastewaters are finally disposed of in most regions of India is either by discharging them to public sewers in urban areas where sewers exist or by disposing them on land (for local gardening and/or soakage underground).

Sometimes, the wastewater is treated, as explained earlier, to make it fit for reuse and thus reduce freshwater intake.

3.4 SUMMARY OF RECOMMENDATIONS

In summary, what this Section teaches us besides rainwater harvesting methods, is the need for better management of our water resources. Mumbai is a very good example of rainwater harvesting since all the rain that falls in the catchment areas every monsoon is held in the different lakes for use by the city. This is, indeed, rainwater harvesting on a large scale. However, it is followed by poor management and loss of water on an equally large scale.

Mumbai should undertake new schemes on a timely basis for adding to the water supply of the city and reduce the relatively large leakage/wastage of water.

> Nearly 35% of the "harvested water" is lost from the lakes by natural evaporation and another 25–30% is lost in conveyance to the city as leakage and "unaccounted" water loss. Thus, a total of nearly 65% (or even more) is just lost. Thus, from a full lake we actually use only $1/3^{rd}$ of it.

Activists who are against rainwater harvesting by the general public point out that it denotes a "failure" on the part of the local municipal body for the following reasons:

- ➢ If less water is lost through better management of the city's supply, rainwater harvesting by the public may not become such an urgent necessity after all.
- ➢ Rainwater harvesting introduces a "dual quality" supply to the city, one controlled by the local authority and the other by different members of the public. Moreover, the upper layers of soil are known to be already polluted from "leaking" sewers where the rainwater may be held for later use. This constitutes a definite health hazard and in case of a problem responsibility cannot be fixed.
- ➢ Experience would warn us that once rainwater harvesting is established, dual-quality supply will become a permanent feature of water supply in the city. It may not remain a stop-gap arrangement for a few years only.
- ➢ One could say that the use of water tankers in some cities today gives more or less the same health hazard! The correct way would be to eliminate all health hazards and place the responsibility squarely on one body (the municipal authority that collects the taxes) to supply necessary quantity and quality of water.
- ➢ In all this, it would be a great help if the water supply service was charged at the right price to the public so that priorities would fall in place automatically, and conservation would be done with greater care.

Rainwater harvesting for the larger metropolitan cities, indeed, remains a question whether it should be done at all. There is a differing school of thought that says our policies need some reconsideration in view of the problems stated.

> Rainwater harvesting, however, appears to be most beneficial for smaller towns, *rural* areas, villages, and the like where public water supplies are not yet fully established and people's needs of water for drinking and agriculture have to be met.

Policies regarding agricultural use of water have also to be reviewed. After all, agriculture is the main user of our national water resource because more than 80% of our available water is used in agriculture alone. We need more water (through rainwater harvesting and groundwater recharge) to generate more income for the farmers while conserving our national water resources.

Observation: Immense benefits are foreseen from adopting better water resources management policies. The use of rainwater harvesting in urban areas as a "fire-fighting" weapon to meet shortages seems undesirable but tolerable under the circumstances for short periods.

Rainwater harvesting seems to be more meaningful in *rural* areas. If practiced more in rural areas, the farmers would benefit from water being available for longer periods, the nation would have more food to eat (presumably at lower prices) and the face of India would change and make us all more prosperous. What an enormous potential exists for us.

REFERENCES

Arceivala, S J, *Planning Rainwater Harvesting Systems*, Rotary Club of Bombay Hills South, Mumbai District 3140.

EXERCISE QUESTIONS

1. How would you spread the message country-wide that rainwater harvesting is a risky business for cities, whereas it is most recommended for rural supplies?

PROJECT/ASSIGNMENT TOPICS

1. Briefly discuss each of the following options for implementation, marking those which you think are more feasible to adopt, and those which might possibly have undesirable side-effects.

   Soft Options

   - Encourage cities to adopt rainwater harvesting to meet shortfalls in supply.
   - Hold seminars on methods of rainwater harvesting.
   - Take steps to fight likely increase in mosquito nuisance.
   - Refer to rainwater harvesting history from ancient times.

Hard Options

- Implement a program to provide rainwater harvesting in rural areas.
- Implement a program to train the trainers in methods of rainwater harvesting.
- Promote development of consultancy and well-drilling services.
- Use ground water to supplement directly harvested rainwater.
- Reuse water to the extent possible.

4 CHAPTER

Preventive Environmental Management

4.1 The New Management Challenges

4.2 Modern Trends in Environmental Management

4.3 Some Issues in Town-planning and Infrastructure Development

4.4 Environmental Impact Assessment

4.5 Case Study: The EIA of Water Resources Development Project

4.6 Life Cycle Assessment (LCA)

4.7 Case Studies of LCA Applications

No one obtains "good environment" as a by-product or in "buy one – get one free" sale offer! Always, it is the outcome of responsible and judicious consumption of resources at our disposal as well as careful planning and management of the consequences of our production and consumption.

As you can guess, it is not easy to define: *What is environmental management?* During the past five decades, industry undertook a variety of activities that changed the face of manufacturing, packaging, advertising, marketing, storing, transporting, retailing, and consuming a variety of goods. Worrying about the so-called "environmental consequences or impacts" is a rather recent phenomenon. All the stakeholders have an opinion these days on environmental implications of the production-consumption activities.

A lot can be written and reported about the efforts by industry, which eventually helped minimize the "environmental footprint" of industrial activities. It would, however, be more significant and important to include all stakeholders and polluters and trace the locus of international thought in this field. It is hoped that such analysis will throw more light on the enabling mechanism evolving worldwide. That process was, in eighties, referred to as "environmental management" and recently christened as "integrated management" or in some other context as "total quality management" and frankly by many many names to suit the setting and perspective and the manufacturing-consuming system.

4.1 THE NEW MANAGEMENT CHALLENGES

Working towards improving the quality of life on earth has always been one of the principal obsessions of people around the world. The tendency to evolve a mix of approaches including regulatory, fiscal and technological fixes has been quite explicit in the context of industrial pollution management. Location-specific, national, regional and global regimes have only been growing in significance over the last few years.

The felt need to define "preventive environmental management" approaches and remedial action in conjunction with each other is of greater significance. Several initiatives have emerged to combat this onslaught on the environment and on the overall quality of life. Laws were put forward in different parts of the world to enforce environmental compliance and remedial action. This trend is increasingly evident in India too.

It is important to recognize that the majority of such interventions have sought to control the activities as a reactive mechanism of governance. However a perceptible shift towards a pro-active preventive framework has also been occurring. The emerging paradigm is "Preventive Environmental Management" (PEM) that integrates Cleaner Production (CP) and remedial action.

The relevance of CP was best articulated in 1994 by the United Nations Environmental Program (UNEP): "If sustainable development is to be achieved, production processes, products, and services have to be reoriented towards new patterns, in order to both alleviate environmental stress and bring better industrial productivity. This requires the development and use of new policy and management tools in both Government and industry, as well as the development and use of environmentally sound technologies with a focus on cleaner and safer technologies that prevent pollution and make efficient use of raw materials".

Work on environmentally sound technologies (1989) has been a forerunner to these CP initiatives which have been growing in diversity and scale over the last eight years providing hope for a potential transition to a cleaner environment.

The preparedness of the business community to also enable this transition is essential because of the following:

- By 2040, reductions of over 90% in material throughput, energy use and environmental degradation will be required in the industrialized world, to meet the needs of a growing world population fairly within the planet's ecological means.
- There is currently no business perspective on the extra 3 billion people that the world is likely to hold by 2025. Rather, every marketing feeler is targeted at the rich 1 billion, living mostly in the developed regions.
- Businesses which have prepared for these revolutionary changes will not only avoid extinction, but will thrive on the rapidly growing demands for eco-efficient goods and services. Companies with Total Quality Management (TQM) as well as with far reaching pollution prevention and waste minimization programmes will thrive.
- Managers should take care to include five key elements of success—a vision, skills, incentives, financial and technological

resources, and a targeted action plan – in the design and implementation of strategies to manage the complexities of change towards eco-efficiency.

- Absolute targets for eco-efficiency are not possible to lay down. This means that companies need to devise dynamic systems for attaining continuous improvement in environmental performance as the targets are updated constantly in the light of new evidence and opportunities.
- While some of the seeds of eco-efficient business are already in evidence, many of today's ways of doing business will be challenged and replaced.

This clarion call to industries for adopting cleaner approaches is also to complement legislation on environmental protection in several parts of the world, prompting them to clean up. In this process, ways of internalizing the negative environmental externalities generated could be devised. The Responsible Care Programme is one such initiative. Eco Industrial Networking (EIN) involving several industries within the industrial estates has also been gaining ground recently. Trade related environmental dynamics is also gaining importance.

The interplay of market influences and environmental imperatives will help understand the economic sense of becoming cleaner. Several firms across the globe have also recognized the importance of factoring the socio-economic dimension of becoming cleaner. While examples of such firms are encouraging, they appear to be only few and far between.

There is scope of optimism in the Indian context. The Air Pollution Prevention and Control Act, the Water Pollution Prevention and Control Act, the Environmental Protection Act, the Public Liability Insurance Act, Minimum National Standards (MINAS), etc., all reflect the intent of environmental governance in detail. Some of the best capacity building initiatives have recently focused on cleaner production through rigorous institutional mechanisms across the country. They are best represented by the elaborate programmes of the National Cleaner Production Centre (NCPC), New Delhi.

The Green Rating Project coordinated by the Centre for Science and Environment has been a classic example. A large number of public interest litigations across the country seeking protection and redressal from the negative externalities tend to destroy the old image of a pollution-tolerant country.

In the international arena, for example, India has apparently played a very prominent role with reference to the Montreal Protocol, devising elaborate regulatory, fiscal and capacity building frameworks to achieve desired results. The trial activities implemented jointly with projects under the United Nations Framework Convention on Climate Change (UNFCCC) are some good examples. Learnings from these pilot projects will guide the development of appropriate interventions to ensure that India's sustainable development imperatives are catered to and not compromised.

While all these aspects of environmental progress are indeed encouraging, a critical analysis of the scale of outreach achieved in sensitizing industry to become cleaner, shows that a great deal is yet to achieved. This is reflected in the persistent acrimony between industry related pollution impacts suffered by the people in the vicinity of firms and downstream areas.

The most recent World Trade Organization (WTO) stand that environmental parameters will be integral part of trade-related compliance re-emphasizes the need to become cleaner. Accordingly the need to achieve higher levels of efficiency in environmental performance cannot therefore be ignored anymore by the Indian industry. A brief review of the range and intensity of environmental chaos experienced by the affected people reveals that much more has to be done in our country to finish India's "unfinished agenda".

It is increasingly recognized that it makes better economic sense to become cleaner! Some significant examples can be cited from India—for others to follow.

Learnings generated from our own country and from other parts of the world need to be integrated with the development of proactive strategies, that link environmental protection with improved economic returns through cleaner production. The significance of relevant tools and techniques has been discussed to serve the bottom line of capacity building *viz.* improving preparedness to act through information support.

India has been on a steep learning curve over the past three decades; more so during the last decade adapting liberalization and evolving her own structures of competitiveness and environmental governance. The domestic and global demand for an environmentally efficient society has been the important driver for transitioning to better technologies and improved production practices.

4.2 MODERN TRENDS IN ENVIRONMENTAL MANAGEMENT

Preventive Environmental Management (PEM) stands for everything that was covered by the traditional "Environmental Management". In addition, it incorporates a dynamic synergy of tools and approaches, which can be easily applied by diverse stakeholders within the society. For instance, a PEM expression by industry can be in the forms of:

- cleaner production
- waste minimization
- value addition to wastes, and
- reducing recalcitrance of wastes.

This could also have significant bearing on improving safety and health at the work place and the surroundings. Technocrats active in areas of technology development can analyze gaps in technologies and production practices to first achieve waste minimization and recycling of resources. Engineers and scientists addressing issues of resource conservation and advanced management of wastes can devise pathways for technical improvement. They could thus help in sectoral transitions to cleaner production with significantly lesser material and energy intensitivites. Thus, it is important to understand the benefits of employing holistic approaches PEM stands for, in diagnosing and interpreting pathways for enhancing efficiencies in production and consumption.

Table 4.1 shows the four critical tools routinely used by modern environmental managers. Under each of these four tools a few sub-topics have been listed, not described. These sub-topics are really discussion topics which can be used for discussion purposes to give a better understanding of the subject.

Table 4.1 *Environmental management tools*

1. Environmental Audit	
1.1	Significance of Environmental Audit as a Tool
1.2	Why Different Audits?
1.3	Protocols and Procedures
2. Environmental Impact Assessment	
2.1	Significance of EIA as a Tool
2.2	EIA is Mandatory for all Development Projects

(*Contd.*)

(Contd.)

2.3	Protocols and Procedures
2.4	A Long Way to Go Before EIAs Facilitate Sustainablty
3. Life Cycle Assessment	
3.1	What is LCA?
3.2	Interpretation of Environmental Attributes
3.3	Use of LCA in Decision Making
4. Environmental Management Systems (EMS)	
4.1	What are "Wastes"?
4.2	Significance of Waste Inventory as a Tool
4.3	Protocols and Procedure of Inventorization
4.4	Significance of EMS as a Tool
4.5	Components of an Environment Management Systems
4.6	ISO 14001: A Model Environment Management System
4.7	Issues Related to ISO Certification

Table 4.2 shows the potential initiatives that may be taken by various stakeholders. These initiatives combine one or many of the tools routinely used by the modern environmental managers described in the above Table. Again, the sub-topics listed under each major initiative are meant for discussion purposes to improve one's understanding of the subject. In some cases a reference to the internet is recommended.

Table 4.2 *Some initiatives and interventions management initiatives*

1. Identification of Potential Interventions	
1.1	What is Cleaner Production?
1.2	Cleaner Production Techniques
1.3	Changes in Indian Industries
1.4	An Illustrated Approach for Cleaner Production of Pesticide Active Ingredient
1.5	Alternate Approaches to Prioritizing Options for Pollution Prevention
2. Greener Chemistry and Cleaner Technologies	
2.1	Green Chemistry
2.2	Green Redox Technologies
2.3	Green Catalyst
2.4	Bio Catalysts
2.5	Phase Transfer Catalyst
2.6	Membrane Separation

(Contd.)

(Contd.)

3. Greener Choices in Process Industry	
3.1	Choice of Eco-friendly Solvents
3.2	Choice of Eco-friendly Reactants
3.3	Choice of Eco-friendly Bio-Fuels
3.4	Choice of Eco-friendly Fuel Cells
4. Redesigning Unit Operations and Unit Processes	
4.1	Mass Exchange Networks
4.2	Pinch Technique
5. Recycle and Reuse of Wastewater	
5.1	Status of Regulation and Compliance In India
5.2	What are Our Options?
5.3	Framework for Sustainable Urban Wastewater Management
5.4	Issues in Water Recycling and Reuse
6. Sustainable and Intelligent Consumption	
6.1	What is Sustainable Consumption?
6.2	Transitioning to Sustainable Consumption in the Context of Globalization
6.3	Why Consumption Pattern Matters?
6.4	Rebound Effects of Consumption
7. Extended Producer Responsibility (EPR)	
7.1	What is Extended Producer Responsibility?
7.2	The Lindhqvist Model for EPR Implementation
7.3	"Design for Environment" as the Ultimate EPR-goal
7.4	Business Strategies Inspired by EPR
8. Corporate Responsibility for Environmental Protection (CREP)	
8.1	What is CREP?
8.2	The CPCB Model for CREP Implementation: Charter for CREP
8.3	Business Strategies Inspired by CREP
9. Eco -Industrial Networks	
9.1	The Concept of Eco-Industrial Development
9.2	Risks and Benefits

As said earlier, It was the great Indian sage, Swami Vivekananda, who perhaps unwittingly set the agenda for all environmentalists as early as 1905 when he said: *"To protect is the voice of the present; to prevent is the divine whisper of the future"*.

Waste Minimization

"Waste minimization" is an important part of this exercise. It consists of examining each step in the manufacturing process for its water requirement and the quality of waste (solid, liquid and gaseous) produced at each step. Where possible, such an exercise may be carried out on an existing factory by its own technical staff, preferably with the help of a process consultant.

Some of the important waste minimization methods available are the following:

1. Good housekeeping (to control wastage from overflows, spillages, unduly high pressures in piping, etc.).
2. Adoption of counter-current flows (to conserve use of chemicals, water, fuels, etc.).
3. Segregation of wastes (to facilitate wastewater treatment and/or reuse).
4. Reuse and recycle (many industries in Mumbai, Gujarat, Rajasthan, Tamil Nadu and other areas reuse sewage after treatment for different purposes (as boiler feed waters, cooling waters, landscape and garden waters, toilet waters, etc.) because often municipal water supplied to industries is expensive compared to treatment costs for reuse. Reliance Petrochemicals at Jamnagar, is a good example where much reuse is done and the balance requirement of water is obtained by distilling sea water which is normally an expensive process. Sometimes, adoption of a "dry" process instead of wet one may help.
5. By-product recovery (Several products, organic and inorganic, are economically recoverable.) For example, recovered biogas may be used as a fuel and help to reduce pay-back period. Another by-product recovery can be chromium from the leather tanning industry.

All the above methods are generally covered under "Cleaner Technologies". Thus, many things have to be done before site finalization. To repeat,

1. First select the manufacturing technology. In doing so, if possible, one with minimum resource requirements of water, power, materials, etc., should be preferred.
2. Then, check availability of required water and other resources required at the tentatively selected site.

3. Whatever the site in question, use "cleaner technologies" which produce less wastes and more easily treatable wastes.

GREEN

The acronym "GREEN" which we use quite freely stands for *"Growth with Resources, Environment Enhancement and Nature"*.

It describes appropriately our basic philosophy, and can be called the basis of our approach to impact assessment. EIA is an orderly and structured method of ensuring that such an approach is followed.

4.3 SOME ISSUES IN TOWN-PLANNING AND INFRASTRUCTURE DEVELOPMENT

We are aware that the governmental town-planning agencies are involved in fixing norms for developments that may occur within their jurisdiction. Their requirements have to be taken into account when carrying out environmental impact studies. The two sometimes may conflict. Some of the salient requirements have been listed to highlight issues in town planning:

1. In the official Development Plan, certain areas may be earmarked for industrial development. Road widths may be prescribed. Certain water bodies, lakes, ponds, canals, etc., may be shown and their use and water quality may be envisaged differently from those assumed in the EIA.
2. The "Floor Space Index" (FSI) for an area in question may be prescribed and not allowed to be exceeded because the public infrastructure may be designed to take care of a certain population density in the area. This infrastructure may include water supply distribution, wastewater collection and disposal, solid waste collection and disposal, etc. If the FSI is changed, the population density may change and the infrastructure may become inadequate to serve the people. EIA must respect the intended FSI.
3. In preparing a "Development Plan" for an area, adequate spaces have to be left for various infrastructure items such as roads, intersections, utilities such as water treatment plants, pumping stations, wastewater treatment plants, solid waste dumping sites, hazardous waste secure landfill sites, open spaces, etc. Either inadequate space or space at a wrong location may be allowed

in a town plan and may either need to be complied with or long proceedings started for amending the plans. This has to be done one way or another.

4. Rehabilitation of Project Affected People (PAPs) is always a tricky question fraught with political undertones. Their accommodation must be given priority and some sympathy, if a project is to proceed peacefully, and meet governmental or funding agency requirements.
5. Once an area starts to get industrialized, unwanted "induced growth" also starts to take place within that area. The effect of unwanted "induced growth" can indeed be enormous. An induced growth is an unplanned residential growth, similar to a slum-like growth, populated generally by poor people looking for employment in the upcoming industries. Their presence leads to a further secondary growth of people involved in rendering services to them such as the ubiquitous "chai-wala", "ricksha-wala", etc. It is not easy to predict this growth quantitatively, but it can be safely said that induced growth is difficult to prevent and can have a greater socio-economic effect in the long run than expected.
6. Similarly, once an area starts to get industrialized, traffic starts to build up. Roads may prove inadequate. Cross-roads may become traffic snarl points. Parking may become obstructive. A whole lot of ills may follow. The estimation of increase in traffic with development of an area remains a weak link in town-planning.

4.3.1 Factors in Site Selection for New Industrial and Civic Developments

Everybody is affected by development. One is either a developer or a beneficiary or a "project affected person" displaced from his habitat. Before any new large industry is established or any new large-scale residential or commercial development takes place a few essential environmental aspects will have to be first considered. Normally, the following steps are taken when any large industry or civic development is to be proceeded with although sometimes some short-circuiting takes place:

- A tentative project proposal is first prepared to consider and understand its various funding, economic and other feasibility aspects.
- A formal, full-scale project proposal is then prepared after incorporating the latest design practices, special requirements to

qualify for any concessions, and after consulting experts/industry organizations, etc., on resources required, processes incorporated, labour, etc.

➢ Select a tentative site (or one or two other tentative and available sites) and carry out "due diligence" studies to choose best site(s).
➢ Undertake an Environmental Impact Analysis (EIA) study based on the selected site and secure approval for site and project.
➢ Proceed with the project.

Sometimes, instead of a project to be developed a product may be desired to be manufactured and marketed. If so, the EIA has to be further followed by a Life Cycle Analysis (LCA) of the likely environmental impacts due to its usage and final disposal.

A Change in Mindset is First Needed

It is important to remember that what is stated below applies equally well to both the public and private sectors. Both can improve an environment or ruin it. The long-term effects in both cases may not become so apparent in the initial years, but the detrimental effects on the ecosystem may nonetheless exist and show up gradually at which time it may be too late to do anything about it.

A new industry may produce a hazardous waste whose presence may not affect anyone in the initial years but gradually ruin a soil or a whole ecosystem through release of its toxic contents. A new road or railway line or some such development by a public authority may seem like a boon initially, but gradually it may lead to destruction of a forest or heritage site because concomitant protective measures were not envisioned. Wrongly located civic developments can cause problems no less than wrongly located industries.

Gone are the days when new factories or developments could be set up at any site we fancied for family reasons. Gone are the days when we could go overseas and buy any old technology we fancied because it was cheaper in initial cost. Gone are the days when we just did "end-of-pipe" treatment of our wastes, whatever they were.

Now, in the modern scheme of things, we have to take a holistic view of all projects whether they are only of a civic nature or involve the manufacture of a commercial product.

Only in this way all the wastes (solid, liquid and gaseous) can be reduced as far as possible first in the manufacturing step and then in all the remaining steps so that they are fully reduced in volume and venom and are treated properly and the effluent reused as far as possible so that net minimum resources are used up and minimum residues go out to the environment. This is what we call "cleaner technologies" in production, operation and disposal.

One would also consult specialists and the concerned Industry associations to learn of the latest thinking in the industry concerning manufacture, waste disposal and environmental matters. Undertake due diligence studies at tentatively selected sites.

Due Diligence Studies

Before confirming a tentatively selected site, carry out "due diligence" studies at site so as to ensure that the site itself does not have problems (often problems inherited from previous owners of the land who might have released hazardous wastes of their own into the soil). Sometimes the land itself may have unwanted toxic material or radioactive deposits naturally available at the site which would hinder subsequent activities. An elementary investigation is necessary to avoid later problems. Typically, due diligence studies were not taken seriously in India until recent times. Lately, however, there are more instances of commissioning those studies when changes in land ownership take place. Slowly, the wisdom in conducting due diligence studies is being appreciated by commercial establishments in metros and industrial locations in India.

4.4 ENVIRONMENTAL IMPACT ASSESSMENT

> "When you can never be too sure,
> prevention is better than cure!"
>
> Anonymous

Environmental impact assessment (EIA) must have been promoted by someone who strongly believed in the above couplet. More rightly, it was promoted by very sensible people who knew that nobody knows exactly how an area will develop, but everybody has a good idea what will happen when some development takes place! There will be many viewpoints, many ideas, perhaps many short-comings too. There will be some good results and some not-so-good results of the development taking place. So, we must have a way of viewing them before they occur. Mischievously said, EIA is a way of worrying before an event occurs!

There is another important thought that EIA cautions us to "look before you leap"! There is a warning that "we have only borrowed the earth from our children who will use it after us". This is said better in a Scottish phrase:

> A little of the earth belongs to those who are living,
> A little to those who are dead,
> But a large part belongs to those who are yet to be born!

So, we have no right to misuse it. Nor do we have the right to use up all our resources as if there will be no tomorrow! The world has not to be used on a "first come, first served" basis. We must use our resources like water, air, land, forests, etc., with care and respect nature in the way we do things.

As EIA procedures have an "advance warning" capability many governments have made EIA studies mandatory before new developments take place, comprising at least the following type of activities:

(a) Civic developments (Motorways, airports, railways, harbours, inland waterways, dams, irrigation projects, townships, touristic developments, etc.)

(b) Industries (Mining, refining, power generation, manufacturing, chemical, rubber, food, hospitality, etc.) and authorities have also specified the contents and other procedures to be followed for making such studies.

Average time for a rapid EIA is about 4–6 months. Much longer period is normally allowed, however, for the so-called "comprehensive EIA"— 15–18 months. A year often has to be allowed so that all the seasons are covered so as to secure enough data for proper study, often using modeling techniques.

The first step required is *"Scoping"* in which the Government, the developer and the public get together to decide on the scope of the report. This eliminates useless information and often focuses the study on relevant and essential matters for the site in question. The length of the report is kept at less than 150 pages (as far as possible) and use of technical jargon is avoided so that the general public can participate more effectively in the later evaluation and negotiations phases.

4.4.1 The Essential Components of EIA Studies

Generally, a full-scale EIA study involves the following essential components and details and is clearly a multi-disciplinary activity. Table 4.3 depicts the typical coverage of various environmental sub-systems and the corresponding specialist to be recruited in the team designated to conduct EIA of a given development project.

Table 4.3 *Listing of the typical environmental sub-systems to be addressed in EIA study and the corresponding specialists to be recruited in the EIA-team*

Sr	Component	Typical Items Covered	Specialists Involved
1.	Air	Wind direction and speed, temperature, precipitation, air quality, noise	Environmental engineer, meteorologist, noise expert
2.	Water	Surface waters, ground-waters, drainage, floods, sedimentation, water quality	Hydrologist, civil engineer, environmental engineer, chemist
3.	Geology and soils	Erosion, slope stability, bearing capacity, soil structure, tectonic activity, land	Environmental/civil engineer, soil expert, agronomist
4.	Biota	Species inventory, species diversity, environmentally sensitive areas, wetlands, deforestation	Ecologist/biologist, forester, environmental engineer
5.	Economy	Resources, demography, land use, employment	Economist, urban planner, environmental engineer
6.	Sociology	Social infrastructure, aesthetics	Urban planner, environmental engineer

The reasons for undertaking a Due Diligence and EIA studies are quite rational. Those can be summarized as follows:

1. It forces one to visualize all environmental consequences before a project is undertaken so that one is sure of the project and the site from the environment point of view. Undertaking only a techno-economic assessment is not enough. Social aspects have also to be considered.
2. It ensures that appropriate technologies are used from the start of the project and nature and environment are respected. The latest "Cleaner Technologies" have to be used in planning, manufacture and waste treatment.

3. It ensures minimum amount of wastes are produced by following the mantra of "Reduce-Recycle-Reuse, as far as possible reaching "ZERO" discharge from the industry, if possible.
4. It ensures that the requirements prescribed by the authorities for availing of "Tax Benefits" or other privileges are possible to meet.
5. Any disasters that could occur at a site are envisaged and mitigative measures built-in. Also, appropriate Disaster Management Plans are drawn up in advance.
6. Export requirements are fully visualized and incorporated into the project from the start. Some countries have rigorous requirements to be met by the products exported to them.

4.5 CASE STUDY: THE EIA OF A WATER RESOURCES DEVELOPMENT PROJECT

As an example, let us take a water resources development project involving construction of a dam, a reservoir and an irrigation system downstream. Such a project can have very wide and far reaching consequences on the socio-economy of the region and the people's health. The proponents of the project (usually government) may mainly focus on the irrigation potential and the economic benefits of the same while political agitation by pressure groups, NGOs, etc. may be related to aspects like:

(a) Displacement of people
(b) Loss of agricultural land, and
(c) Submergence of historical sites, if any

EIA studies would have to look at all aspects and generate answers to anticipated queries. This is no mean task. In a system where pollutants can spread far and wide through rivers, soils, food chains and webs, and can have "boomerang" effects on health, the ecosystem and the economy. Some assessment of their likely environmental consequences has to be made before their discharge is permitted at a site. To make this assessment in an orderly manner, the various proposed "actions" or project activities and the nature of their likely impacts (both during the initial construction phase and subsequently during the operation phase) have to be listed carefully. For example, a suggested outline for the above type of project would have to include studies on at least the following (see Table 4.4).

Table 4.4 *Recommendations for inclusion of the various proposed "actions" or "project activities" and the nature of their likely impacts in a water resources development project involving construction of a dam, a reservoir and an irrigation system downstream*

The Dam and Spillway	
1.	Hydrological aspects (yield, flood, etc.)
2.	Remote sensing of geo-structural conditions
3.	Seismic effects
4.	Dam and spillway type and height
5.	Safety considerations
The Reservoir and its Catchment	
6.	Submergence effects; population displacement
7.	Land use changes; deforestation
8.	Pollution potential in catchment area from drainage, habitation, industries, air pollution, etc.
9.	Water quality changes due to impoundment, recreational use
10.	Mosquito breeding potential
11.	Loss of cultural / historical assets
Irrigation Command Area	
12.	Allocation and distribution of water
13.	Agricultural productivity
14.	Soil-water-crop-climate relationship
15.	Water logging potential
16.	Effects on groundwater
17.	Soil drainage and salinity build up potential
Downstream River	
18.	Potential quality and quantity changes
19.	Effects on beneficial uses of water such as navigation, power generation, agriculture, fisheries, etc.
20.	Erosion, deposition effects
21.	Salt water intrusion potential
Ecosystem and Health Hazards	
22.	Effects on aquatic ecosystem
23.	Disease vectors
24.	Impacts on vegetation and forests
25.	Impacts on wildlife
26.	Special needs of any rare and endangered species

To be able to obtain the right data and interpret it correctly, a multi-disciplinary team approach is essential, otherwise the answers may only be "no effect anticipated". Of course, certain socio-economic parameters are subjective in nature and could lead to controversies.

To facilitate EIA work, a matrix is often developed for each project listing several "Project Activities" against specific "Environmental Impacts" likely. A typical example of an industry is given below with a few items only.

The reader is expected to prepare a full-scale matrix to complete the example. "Project Activities" may include items like the following from the cradle to the grave:

(a) Stack emissions
(b) Cooling tower blow-down
(c) Effluent from treatment plant, etc.

Against each of these activities the various likely "environmental impacts" at the site need to be listed such as the following:

(a) Likely Environmental Impacts
(b) Water Quality downstream in river
(c) Air Quality downwind
(d) Noise level, etc.

Thus the "Project Activities" and the "Environmental Impacts" form a kind of matrix and the likely impacts in each case (after control measures are taken) can be noted in a short and coded form in the corresponding boxes in the matrix, for ready reference, using the following codes:

Nature of impact: Beneficial (+), Adverse (–) (These signs are not additive)

Magnitude: Slight (1), Moderate (2), Appreciable (3)

Duration: Permanent (shaded box), Temporary (Unshaded box)

Extent of impact: Local or beyond study area

NA: Not applicable

Those impacts that are appreciable or permanent or extensive and need to be investigated further in detail are marked on the matrix (shaded). Technique such as "mathematical modeling" is used to study them further. For different industries and civic projects an appropriate matrix may need to be developed separately for each case. The EIA report has to explain the likely impacts expected and what sort of measures will be taken to mitigate the impacts.

Public Hearing

After the study is made, it is submitted to Government for approval. An expert committee is usually set up to review the study, ask further questions of the proponent and his team and prepare an assessment synthesizing the various expert comments and opinions. This report is then put up for a public hearing at site.

A public hearing ensures local people are not disadvantaged by a project at a site selected by the proponent. For this reason, a site meeting is mandatorily organized to give the local people an opportunity to speak up in case they have any doubts. Permission to operate at a site may be refused if strong reasons are revealed.

If there is public opposition to a project, the real reasons must be deciphered. Opposition can often be a case of "NIMBY" ("not in my backyard") as people often oppose just because they are afraid development will bring outsiders into an area, and only the outsiders will prosper whereas the local peoples' way of life will be affected adversely. In such a case public protest can take on political overtones and even lead to "Development Hatao" in a region, so to speak.

Finally, after a public hearing at site, the final report is prepared giving reasons for granting or refusing the permission. Compliance requirements are also spelt out, based on which the Government authorities follow up to ensure project requirements are fulfilled. This follow up is crucial.

4.6 LIFE CYCLE ASSESSMENT

As stated earlier, sometimes, instead of a civic or development project, a product may be desired to be manufactured and marketed. If so, the EIA has to be further followed by an analysis of the likely environmental impacts due to its usage and final disposal. This extended EIA is referred to as Life Cycle Assessment (LCA).

In a limited sense, LCA can be viewed as an extended EIA carried out to include the period after a product is manufactured. The period up to and including manufacture is covered by EIA. The period after its manufacture and marketing is covered by a post-EIA study until the useful life of the product comes to an end and the same is disposed off either with municipal solid waste or otherwise sent for separate handling as a special, specific or hazardous waste to its final resting place. Hence, a LCA covers the product from what is commonly known as "cradle to the grave".

Because LCA takes the "cradle-to-grave" approach, it differs from conventional tools such as environmental audit or environmental impact assessment (EIA). It is important to remember that LCA takes into account both local as well as global impacts and is hence more suited in the current environmental crises, where we are facing challenges such global warming and oil spills.

LCA has emerged as very useful tool since its inception in 1990's when Society of Environmental Toxicity and Chemistry (SETAC) has advised to conduct LCA for analyzing environmental problems. The inappropriate use of LCA to market the products has put pressure to standardize LCA methodology and led to the development of the LCA standards in the International Standards Organization (ISO) 14000 series (1997 through 2006) LCA considers all three phases viz.:

- construction or manufacturing phase,
- operation or exploitation phase, and
- dismantling or end-of-life phase of the product to be evaluated.

LCA can estimate various impacts on environment such as climate change, stratospheric ozone depletion, tropospheric ozone (smog) creation, eutrophication, acidification, toxicological stress on human health and ecosystems, the depletion of resources, water use, land use, and noise—and others.

Ideally, a complete LCA would include four separate but interrelated components:

Goal Definition and Scoping: Define and describe the product, process or activity. Establish the context in which the assessment is to be made and identify the boundaries and environmental effects to be reviewed for the assessment. In this step functional unit is selected, which characterizes the function of the system under consideration, depending upon the aims of the study.

Inventory Analysis: In this phase, the energy, water and materials (resources) usage and environmental releases (e.g., air emissions, solid waste disposal, waste water discharges) incurred throughout the life cycle of a product, process, or activity will be identified.

Impact Assessment: In this phase, potential impact of environmental releases identified in the inventory analysis on human and ecological effects of energy, water, and material usage are estimated. Different methods have been developed to ease the comparison between the impacts caused by various pollutants. In any given method, impacts calculation includes three steps: characterization, normalization and weighting. Characterization step deals with transforming he inventory flows into impact indicators quantified using specific characterization factors for each of the pollutant. Normalization is by dividing each studied system impact by the corresponding reference impact calculated for a region or country. Weighing is optional and can be used when the decision maker is has any priority for impacts.

Interpretation: In this phase the opportunities for mitigation of the impacts of products or services are identified. This gives insight in product development, improvement in the process efficiency, etc.

The overall principles and framework are described in the ISO 14040: 2006. The framework shows the relations between the different phases as depicted in Fig. 4.1.

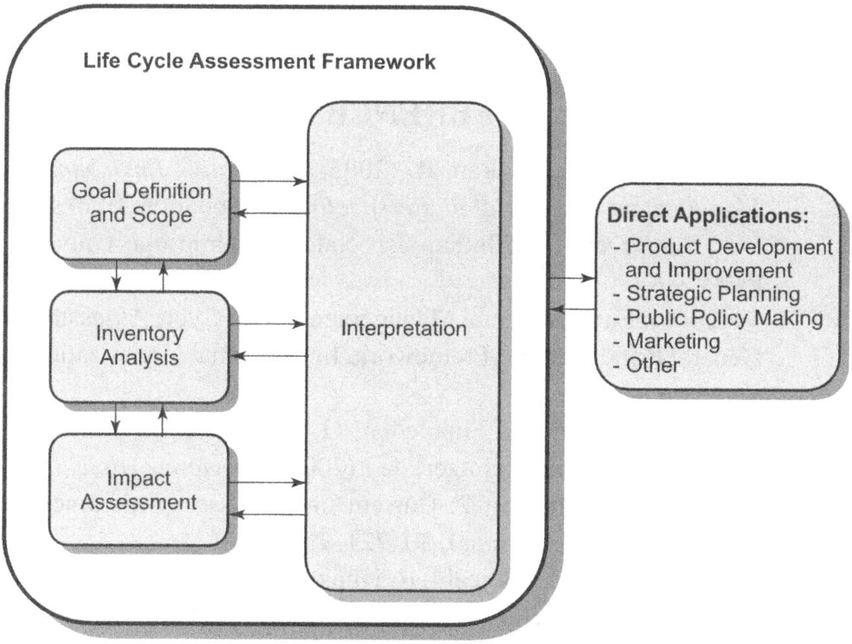

Fig. 4.1 *Phases of LCA as described in ISO 14040 (2006) procedure*

4.7 CASE STUDIES OF LCA APPLICATIONS

Case Study 1: *Incandescent Lamp (IL) v/s Compact Fluorescent Lamp (CFL)*

For analysing such a complex product LCA is very useful tool. The products are studies for complete life cycle and results shows that the CFLs are more environmentally friendly than ILs when the impacts of electricity production are taken into account. More details can be found at http://rmi.org/Knowledge-Center/Library/C08-12_LCAFluorescent IncandescentLamps.

Case Study 2: *Comparison of Packaging Materials*

Due to increased consumption patterns across the borders huge amount of packaging materials are used which finally land up in landfill. LCA provides very good platform to evaluate the different packaging materials from environmental perspective. Many studies have been carried out to study the packaging materials for foods and beverages. More details can be found out at http://www.ifeu.de/english/index.php?bereich=oek.

REFERENCES

Asolekar, S R, and Gopichandran, R, (2005), *Preventive Environmental Management – An Indian Perspective*, Foundation Books Pvt. Ltd., New Delhi (the Indian associate of Cambridge University Press, UK).

ISO 14040 (2006): Environmental Management – Life Cycle Assessment – General Principles and Framework. International Organization for Standardization.

Pennington, D W, Potting, J, Finnveden, G, Lindeijerd, E, Jolliete, O, Rydberga, T, and Rebitzer, G, (2004). *Review article:* "Life Cycle Assessment Part 2: Current impact assessment practice", Environment International, 30, 721–739.

Rebitzer, G, Ekvall, T, Frischknecht, R, Hunkeler, D, Norris, G, Rydberg, T, Schmidt, W P, Suh, S, Weidema, B P, and Pennington, D W (2004). *Review article:* "Life Cycle Assessment Part 1: Framework, goal and scope definition, inventory analysis, and applications", Environment International, 30, 701–720.

Recommended Additional Reading

Arceivala S J, and Asolekar, S R, (2007), *Wastewater Treatment for Pollution Control and Reuse*, Tata McGraw Hill, India.

EXERCISE QUESTIONS

1. Why is it true that starting a new project without an EIA would be a serious folly on the part of the promoter?
2. Explain why site selection has to be done with great care, starting with a due diligence study? Why is a holistic approach necessary to follow in planning a new project?
3. Distinguish between "due diligence", EIA and LCA.

PROJECT/ASSIGNMENT TOPICS

1. Assume any industry you like and prepare a matrix on the above lines, noting the impacts likely to occur and how you propose to mitigate them.
2. As an exercise, prepare an EIA matrix for a transport project in which a new metro line is to be laid between two points in a large city, and incorporating an underground section, an over-ground section, related stations, ingress, etc. Assume required data.
5. Conduct LCA for comparing use of plastc tea cups with clay cups (matkis or kulhad). List the impact categories that you would like to take into consideration for study.
6. Briefly discuss each of the following options for implementation, marking those which you think are more feasible to adopt, and those which might have undesirable side-effects.

Soft Options

- Monitor new projects to see that all approval conditions are implemented.
- Appoint more monitoring staff.
- Levy more stiff fines if implementation is lacking.
- Hold seminars.

Hard Options

- Implement a program to train the trainers for undertaking EIA work.
- Build up expertise from various fields (mathematical modeling, meteorology, remote sensing, various industrial manufactures and export, chemical analyses, environmental engg, etc.).
- Organize site meetings for public discussion.
- Ensure strict follow-up of conditions under which a project is approved for execution.

5 CHAPTER

Air Quality—Ambient and Indoor

5.1 Earth's Atmosphere, Some Air Pollutants and their Health Effects

5.2 Some Concepts from Meteorology Affecting Air Movements

5.3 Industrial Stack Emissions and Control Strategies

5.4 Traffic Pollution and its Control

5.5 Indoor Air Pollution and Ventilation

5.6 Ozone Depletion and Acid Rain

5.7 Noise Pollution

Patriotism is really our love for the air we breathe, the water we drink and the food we eat in our country.

5.1 EARTH'S ATMOSPHERE, SOME AIR POLLUTANTS AND THEIR HEALTH EFFECTS

Our earth is enveloped by air, which we call "atmosphere". It extends up to 600 kilometres from the surface of Earth. The atmosphere supports life on earth and plays an important role in maintaining the geo-hydrological cycle. It absorbs solar radiation and other natural electrical and magnetic forces which might harm the ecosystem and provides a life-supporting moderate climate.

Dry air is composed of various gases, approximately by volume, nitrogen (78%), oxygen (21%), argon (0.9%), and carbon dioxide (0.035%+) as major gases, and neon, helium, hydrogen and ozone in smaller concentrations. Earth's atmosphere also contains about four percent of water in various physical forms such as vapour, liquid and solid. At a given location, air has a more or less constant composition except for water vapour and carbon dioxide. Though availability of water in the upper atmosphere is in traces, it plays an important role in the atmosphere in weather changes and warming of the atmosphere by absorbing solar energy.

The atmosphere is divided into layers as troposphere, stratosphere, mesosphere and ionosphere (see Fig. 5.1). Properties of the atmosphere change in the various atmospheric layers. We are mainly concerned with the first layer which is very important for supporting life on earth although life is affected by many factors besides the atmosphere.

Various functions performed by the earth's atmosphere include the following:

- ➢ Allow vitally important life-supporting sunlight to earth.
- ➢ Maintain the earth's mean temperature for sustaining life.
- ➢ Safeguard life from the harmful ultra-violate and other radiations of sun.
- ➢ Enable movement of energy on earth.
- ➢ Support and maintain the hydrological cycle.
- ➢ Recycle and store nitrogen, oxygen and carbon dioxide in the ecosystem.

- Dilute and transform various natural and manmade emissions.
- Protect from smaller celestial objects (meteors) entering the atmosphere.

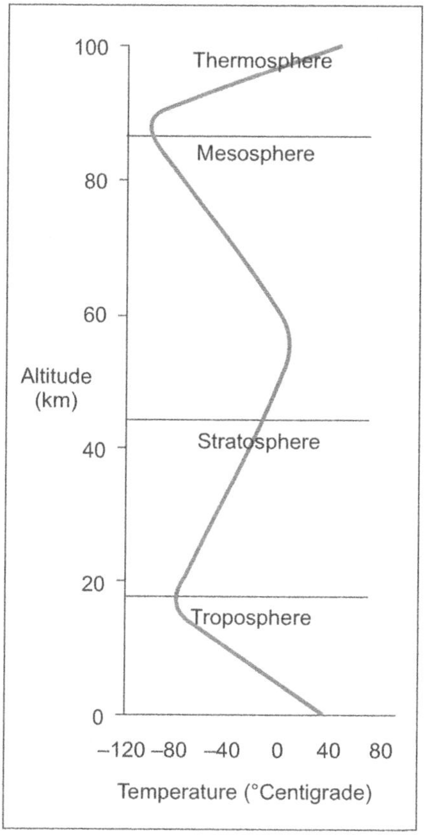

Fig. 5.1 *Layers of Earth's atmosphere. (After Env Poll & Control 4th Ed by Pierce, Weiner and Vaselind Pub: Butterworth-Heinemann (1998))*

5.1.1 Some Air Pollutants, their Sources and Health Effects

Air pollution is caused by the presence of chemicals, particulates or biological substances having potential of harming or creating discomfort to humans, living beings, and or causing damage to the earth's environment. Such substances are termed as air pollutants. Air pollutants can be solid, liquid or gaseous in nature. They may occur indoors or in the outdoor environment. Their concentration and length of exposure are important in understanding their effects. They may affect in the present or in the future or both, and their effects can extend locally, regionally or globally.

The harm due to air pollutants can vary over a wide spectrum from common respiratory ailments to damage to the flora and fauna, plant life and other elements of ecosystem. Sometimes, the ailments may be of a serious and irreversible nature such as cancer.

Sources of air pollution can be differentiated depending on their origin as given below.

Manmade sources (anthropogenic)

1. Domestic heating in colder months
2. Industrial manufacturing and power generation plants
3. Transport vehicles, trains and airplanes, and
4. Waste disposal plants

Manmade sources of air pollution can generally be estimated and, in most cases, controlled effectively.

Natural sources Examples of natural sources are volcanoes, dust storms, etc. It is impossible to estimate the amount of ash or dust or various gases coming out of a natural eruption.

Sources of air pollution can also be classified according to their mobility.

- ➢ Mobile sources include all types of communication and transport systems for land, water or air movement. Automobiles have an increasing impact owing to their fast increasing numbers mainly in urban areas and their pollution depends on the type of fuel used. Aircraft use fine fuel and discharge their pollutants not only in greater quantity but also at high altitudes where they go unnoticed. Ships generally use heavy oil and discharge much polluting substances at high seas where they also go relatively unnoticed.
- ➢ Stationary sources may be point sources or area sources and include the burning of fuel from residential and industrial areas, solid waste incineration plants, industrial stack emissions, and emissions from power plants, etc.

The type of air pollutants discharged to the atmosphere depends upon the fuel used and nature of the activity. Power plants using coal as fuel contribute both particulates as smoke and sulphur oxides, while solid waste incineration plants emit particulate matter and products of incomplete combustion as carbon monoxides.

Some of the important types of air pollutants are:

- ➢ Particulates as suspended particles (SPM) or respirable particulate matter (RSPM)

    - Dust, e.g., from construction activities, traffic
    - Soot, e.g., burning of wastes, fuels, garden wastes
    - Pollen grains from gardens, agriculture
- Aerosols of organic and inorganic origins
    - Polycyclic aromatic hydrocarbons (PAHs), Polychlorinated biphenyls (PCBs),
    - Acid mists
- Smoke
    - Automobile exhausts, e.g., Carbon particles
    - Combustion of kerosene, wood, coal for cooking and heating purposes
    - Exhausts from Industrial processes, e.g., SPM, RSPM
- Gaseous pollutants
    - Gaseous pollutants arising out of various activities, e.g., SO_x, SPM
    - Gaseous products of combustion of fossil fuels, e.g., NO_x, CO, CO_2, HC
    - Gaseous pollutants resulting from industrial processes, e.g., Cl_2, NH_3, etc.

Particulates and Aerosols

The particulate matter present in the air is called suspended particulate matter (SPM). The properties by virtue of which a particle can remain in air are the size of particle and the specific gravity of the particle. Particles of silica or metal of high specific gravity need to be of very small size as compared to fibrous cotton particles of low specific gravity.

Particulate matter of size smaller than 10 μm can enter the human respiratory system while breathing and hence is termed as respirable suspended particulate matter (RSPM). The RSPM has capacity to reach right up to the lungs easily. For this reason, air quality standards emphasize on the presence of particulates of 10 μm (PM 10) and 2.5 μm (PM 2.5) apart from the general term SPM.

Besides size, the effect of particulates on human beings and other elements of the environment depend upon its composition such as organic, inorganic or inert.

Table 5.1 *Different types of particulates with typical particle sizes*

Composition of Particulate	Type of Particulate Matter
Organic	Soot, pollen grains, cotton fibres, diesel particulate matter (DPM)
Inorganic	Industrial dusts, acid mists
Inert	Dust, cement particles, silica, asbestos

Table 5.2 *The sizes of particulates and aerosols*

Coarse Sand	Fine Sand	Spray	Dust	RSPM	Smoke	Gases	Bacteria	Virus
500–1000 µm	125–250 µm	10–70 µm	1–75 µm	>10 µm	0.1–1 µm	>0.001 µm	0.5–5 µm	0.05–0.5 µm

1 mm = 1000 µm

Tables 5.1 and 5.2 Ref "*Env Poll & Control*" 4th Ed by Pierce, Weiner and Vaselind Pub: Butterworth-Heinemann (1998)

The unit of measurements for the particulate matters are represented as mass per unit volume of air. The common unit of measurement is microgram/cubic meter (µg/M^3) or microgram / Normal cubic meter (µg/NM^3).

Gaseous Pollutants

Major air pollutants present in the gaseous form in the atmosphere are the various products of combustion with or without oxygen. Fossil fuels such as hydrocarbons of various types and coal, used in automobiles or in industries as a source of energy, release gases and particulates as the products of chemical reaction . Other industrial and urban activities which contribute the products of anaerobic reactions also enter the atmosphere. Uncontrolled bio-methanation at municipal solid waste dumping grounds, swamps, some industrial solid waste disposal sites and industrial processes along with natural methane contributors as large live stock population on globe are the major sources. Some of the air pollutants are:

- Oxides of Nitrogen (NO_x)
- Oxides of Sulphur (SO_x)
- Carbon mono oxides (CO)
- Carbon dioxide (CO_2)
- Methane (CH_4)
- Ammonia, etc.

Among the above, oxides of nitrogen, nitrogen dioxide (NO) and nitrous oxide (N_2O) are of manmade origin. The rest are formed in nature.

Gasoline in the form of various hydrocarbons such as petrol, diesel or compressed natural gas (CNG) is oxidized in the automobile engines and under ideal conditions gives out CO_2 and water. In practice such reactions do not take place because of the quality of fuel, engine efficiency, air to fuel ratio, fuel injection mechanism and other driving conditions affect the emissions. The combustion results in generation of carbon monoxide, non-reacted HC or fuel, oxides of nitrogen, etc. Additives to the gasoline come out after the reactions in some form as pollutants.

$$C_8H_{18} + O_2 \approx CO_2 + H_2O \quad \text{Ideal reaction}$$

$$\text{Hydrocarbon} \xrightarrow{\text{Oxidizes}} \text{Carbon dioxide} + \text{Water}$$

In real practice the Automobile pollution is resulted as below:

$$C_7H_{16} + \text{Additives as Pb} + O_2 \approx CO_2 + CO + NO_x + SO_x + \text{Unburnt Hydrocarbon} + \text{Lead}$$

$$\text{Gasoline} + \text{Lead (if present)} + O_2 \approx \text{Products of combustion} + \text{Additives as it is} + \text{HC}$$

Biodegradation of solid wastes from land disposal sites, swamps, etc., releases gases like ammonia, methane and hydrogen sulphides after anaerobic reactions. Industrial processes release various types of air pollutants depending upon the raw materials, processes and products.

Problem 5.1: A metropolitan city has vehicle population of two million. Out of this 500,000 are cars. Assuming that all are gasoline (petrol) driven and on any working day, 60% vehicles come on road. An average run of a vehicle is of 20 kilometres, which consumes 1.5 litres of fuel. How much fuel is consumed by the cars on a week day?

Fuel consumed on a week day

$$= (\text{Total cars on road}) \times (\text{Fuel consumed per car})$$
$$= (500{,}000 \times 60\%) \times (1.5 \text{ L/car}) = 450{,}000 \text{ L}$$

The type and condition of vehicle, total vehicle population, type of fuel and driving conditions, these factors are responsible for consumption of fuel and in turn the air pollutants in the urban atmosphere.

The emissions can also be estimated from the Litres of fuel consumed (Chapter 2).

Problem 5.2: A power plant uses 100 Metric Tonne of coal per day for the generation electricity. The coal normally used in the power plant has 1.5% of sulphur as impurity. What will be the emission if the coal degrades with sulphur content of 3%?

Total sulphur burnt in coal per day

= Coal consumed per day × (% Sulphur as impurity)

= (100 MT/day × 1.5 %) = 1.50 MT

$$S + O_2 \approx SO_2$$

32 (1 mole) + 16 × 2 (1 mole) = 64 (2 moles)

One mole of Sulphur combines with one mole of oxygen to give two moles of sulphur dioxide.

If the quality of coal degrades and as given in above problem doubles than the quantity of the sulphur dioxide will be doubled.

Secondary Pollutants

All above are the primary pollutants. When some of the primary pollutants, combine physically, chemically or otherwise under certain atmospheric conditions, they generate secondary pollutants such as photochemical smog or just smog. Smog is formed by combining smoke and fog. Smog is chemical reaction which takes place in presence of solar radiation (UV rays) and hydrocarbon as particulates. Other secondary pollutants are ground level ozone, Peroxyacetyl Nitrate (PAN), etc.

Photochemical reaction during smog formation

Reactions inside the automobile engine

$$N_2 + O_2 \rightarrow 2NO$$

Reactions occurring in the atmosphere

$$2NO + O_2 \rightarrow 2NO_2$$

$$NO_2 \xrightarrow{\text{UV radiation}} NO + O$$

$$O + O_2 \rightarrow O_3$$

$$NO + O_3 \rightarrow NO_2 + O_2$$

$$HC + NO + O_2 \rightarrow NO_2 + PAN$$

Briefly, to recapitulate, we give in Fig. 5.2 the family of air pollutants and a few of the various industries and their operations which are released as air pollutants.

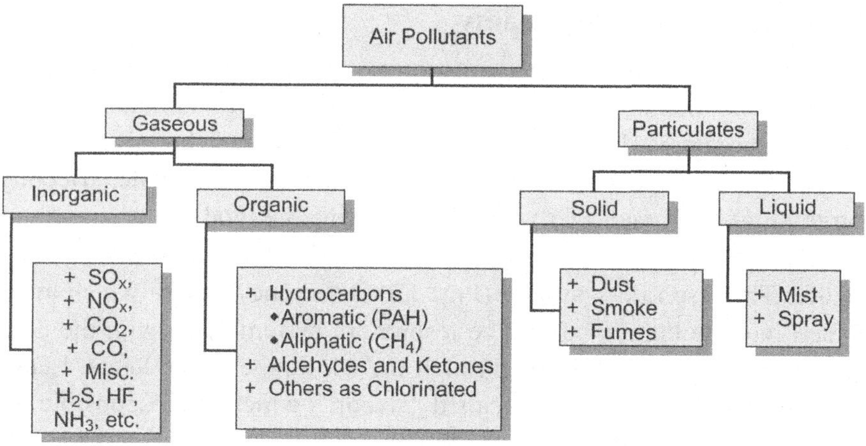

Fig. 5.2 *Air pollutants in brief. (Courtesy: M.N. Rao and H.V.N. Rao, (2007), Air Pollution)*

Table 5.3 *Industrial pollution (Courtesy: "Introduction to Env Eng", (2004, 2010), Vaselind, Morgan & Heine)*

Industry	Source	Nature of Pollution Problem
Primary metals	Blast furnaces, sinters, smelters,	Melting operation gives-off CO, smoke, dust, ash and metallic oxides depending on impurities in ore or scrap used. Smelting gives SO_2, lead, arsenical copper fumes depending on metal smelted
Fabricated metal production	Cupolas, quenching, metal plating	Metallic fumes and dusts from foundries, solvent mists and vapours from coating/finishing, smoke and fumes from quenching, heat treating
Mining and minerals including cement	Materials handling	Dust, fumes, CO, waste explosive gases, smoke from melting and kiln operations. Dusts from mechanical processes
Petroleum refining	Catalysts, regeneration, sludge incinerators	Practically all forms of air pollution arise from refineries, petroleum vapours, SO_2, smoke, dusts, odours, mists from cracking
Pulp and paper	Furnaces, kilns, smelt	Emissions due to combustion, organic solvents, lead oxide from printing works
Textiles	Fibre cleaning and handling	Emissions from heat and power gen. Lint and fines, organic vapours, H_2S
Chemical and allied chemicals	Thermal, acid, grinding	Practically all forms of air pollution, dust, smoke, chemicals, power gen
Thermal power generation	Materials' handling, combustion	Emissions from fuel combustion. Coal dust
Nuclear power generation	Materials' handling	Incineration isotopes, radioactive wastes

5.1.2 Permissible Air Quality

Air pollutants affect all living constituents of earth's environment, man, animals and plants, equally. Air pollutants can affect human health through the air we breathe (indoors or outdoors) and through the food we eat. Pollutants emitted from stacks can spread far and wide, affecting agricultural fields and river waters and entering the food chain.

The quality of air humans breathe has a direct effect on their quality of life. The respiratory system (Fig. 5.3) is the one through which air is routed into the human body. The respiratory system starts with the nose and nasal cavity which leads through the wind pipe to the lungs. Lungs are composed of tiny nodules called "alveoli", which are responsible for transfer of oxygen from the air to the blood. If the air is accompanied by any kind of air pollutant, it affects the human being. These effects can be chronic to acute, i.e., acting very slowly to instantaneous depending upon the type of pollutant, concentration and the duration of exposure.

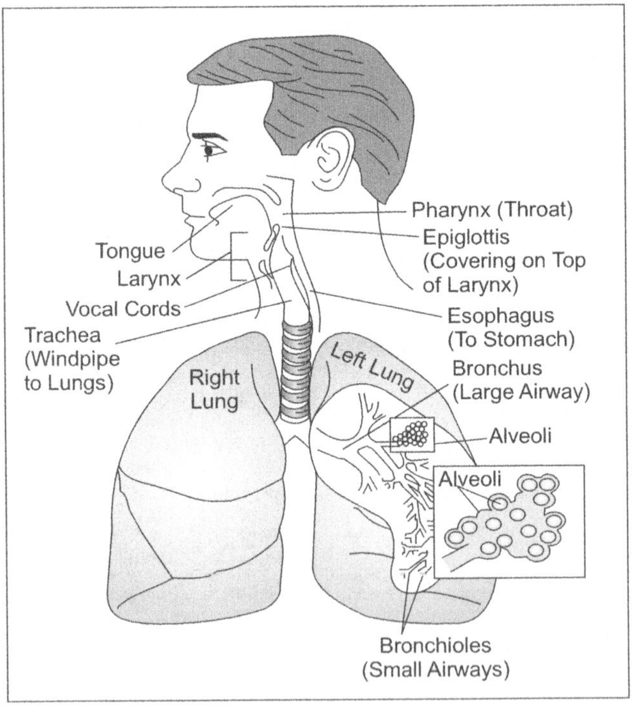

Fig. 5.3 *Human respiratory system. (Courtesy: Department of Medical Surgical Nursing, The Ohio State University Medical Center)*

The following are some of the air borne infections of importance in daily life:

1. Upper and lower respiratory tract infections (bacterial, viral, etc.)
2. Chronic Obstructive Pulmonary Diseases (COPD) such as chronic bronchitis
3. Silicosis (Irreversible lung damage caused by silica dust)
4. Siderosis (Irreversible lung damage caused by dust rich in fine iron particles)
5. Asbestosis (Irreversible lung damage caused by asbestos fibrous particles)
6. Bysinosis (Lung damage caused by cotton fibres)

The National Ambient air quality standards as given by the Ministry of Health, Government of India (2009) are given in Table 5.4.

Table 5.4 *National ambient air quality standards (Courtesy: Govt. of India, CPCB notification, New Delhi, 18th Nov. 2009)*

S No.	Pollutant	Concentration in Ambient Air
1.	Sulphur Dioxide (SO_2)	80 μg/M^3
2.	Nitrogen Dioxide (NO_2)	80 μg/M^3
3.	Particulate matter, PM_{10}	100 μg/M^3
4.	Particulate matter, $PM_{2.5}$	60 μg/M^3
5.	Ozone (O_3)*	180 μg/M^3
6.	Lead (Pb)	1.0 μg/M^3
7.	Carbon Monoxide (CO)*	0.4 mg/M^3
8.	Ammonia (NH_3)	400 μg/M^3
9.	Benzene (C_6H_6)#	0.5 μg/M^3

*One hour Time Weighted Average; #Annual TWA; All other with 24 hour TWA

Measurement

The air pollutants are measured on the basis of mass per unit volume of the air (μg/M^3; mg/M^3) or sometimes as parts per million parts of air (ppm).

It often becomes difficult to judge the exact quality of air because it may meet a few of the above standards and not meet some others. Thus, the concept of Air Quality Index (AQI) has been developed in some countries in which the Index is correlated with observed health effects so that the public can be warned that the air is likely to be "acceptable" or is "hazardous". No such AQI has yet been developed for Indian cities.

5.2 SOME CONCEPTS FROM METEOROLOGY AFFECTING AIR MOVEMENTS

Meteorology is the science related to the study of wind, temperature, pressure, solar radiation, precipitation, humidity in the earth's atmosphere. As the movement of air in the atmosphere is affected by the above meteorological parameters, the air pollutants also meet the same fate.

Wind

Earth's atmosphere is dynamic in nature because of air movements. These movements are caused by atmospheric pressure variations, roughness of the earth's surface, and temperature variations.

Presence of gases in the atmosphere causes changes in the atmospheric pressure. The weight of the air at any place is called atmospheric pressure. At sea level the atmospheric pressure is 760 mm of mercury or 1 kg/cm^2. As we ascend the atmosphere, pressure decreases, it is approximately 554 mm of mercury or 0.75 kg/cm^2 at an altitude of 2.6 km above sea level.

Roughness of surface is caused by the topography (surface features) of the terrain such as mountains, valleys, flat land and by the urban structures such as buildings, bridges, etc. This roughness affects the air movements.

Temperature variations are due to the thermal turbulences caused by land and water masses. "Heat island effect" is due to changes in the air temperature due to the presence of urban areas in the midst of other areas.

Air in motion is called as wind. Wind causes dispersion of the air pollutants in the atmosphere. The direction, intensity and patterns of wind are responsible for the dispersion of air pollutants. The wind velocity, pattern and direction at a particular geographical position on time scales are graphically represented by wind rose diagrams. The profile of the wind velocity in the urban area having tall structures is different as compared to the rural area without such structures. Topography of a particular area also affects the dispersion of pollutants.

Gaseous pollutants emanating from a stack (chimney) will move in the direction of wind. If the pollutants are released close to the surface or through small height stack they will affect the neighbourhood as the pollutants will disperse and remain at lower levels. When the pollutants are released at heights through high stacks they will travel a longer distance and thus get diluted depending upon the wind velocity, atmospheric conditions (stable, unstable and neutral, cloudy or sunny). Thus, taller stacks are needed for dispersing larger volumes of emissions from a factory and smaller heights for smaller emissions.

Lapse Rate

Atmospheric temperatures in the lower atmosphere play an important role in the transmission, propagation and dispersion of air pollutants entering the atmosphere. The fate of air pollutants depend on the temperature profiles of the atmosphere. During the day, solar radiations are absorbed by the land mass and the air in contact gets warmer hence becomes lighter and moves vertically, which is replaced by air comparatively cool over the water bodies such as sea. This causes sea to land movement of air during day. At night, land cools faster than water; hence reverse is the flow of air, i.e., from land to sea in the night. This causes turbulences in the air movement.

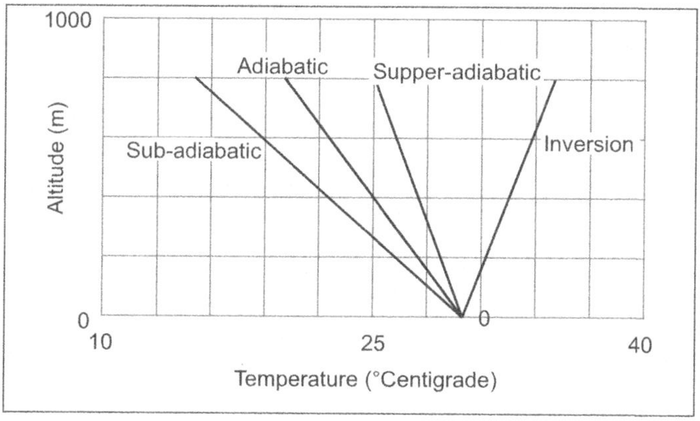

Fig. 5.4 *Lapse rate (Change of Temperature, $\delta T/\delta Z = -1°C/100m$)*

Another reason for air movement is temperature reduction as the altitude increases in the atmosphere. The rate of decrease in temperature with altitude is termed as lapse rate. Normal lapse rate is approximately one degree Centigrade per one hundred meter rise in elevation. This is called the *Adiabatic Lapse Rate*. When the fall in the temperature is less or more than the normal, the transport of the air pollutants is accordingly affected. Such conditions are natural and sometimes manmade too. Figure 5.4 shows graphically the adiabatic, sub-adiabatic and super-adiabatic lapse rates.

Inversion is a condition when the vertical air movement within inversion is almost nil and is called an extreme sub-adiabatic condition.

The inversion conditions are worst conditions from air pollution point of view because pollutants tend to accumulate in a stratum and concentrations build up instead of rising upwards and diluting off.

Variation in the atmospheric pressures is responsible for cyclones and anti-cyclones and can also affect the movement and transport of air pollutants. Similarly, changes in the intensity of solar radiations and changes in humidity also occur due to atmospheric conditions and geographical location on earth. These changes have a bearing on air pollutants.

Meteorological data at any geographical location is used to predict the impact of any human activity on the environment at large. This data is precipitation, wind profiles, ambient temperature, climatic conditions at any time such as day, night or season is used with mathematical models and the monitored concentrations of various pollutants to generate the air quality profile. Any on-going or future man made activity of urban centres, major construction activity, industry or industrial clusters can affect the living and non-living and is quantitatively and qualitatively defined by the Environmental Impact Assessment (EIA). Such air quality models are used for scientifically predicting the immediate and long-term effects.

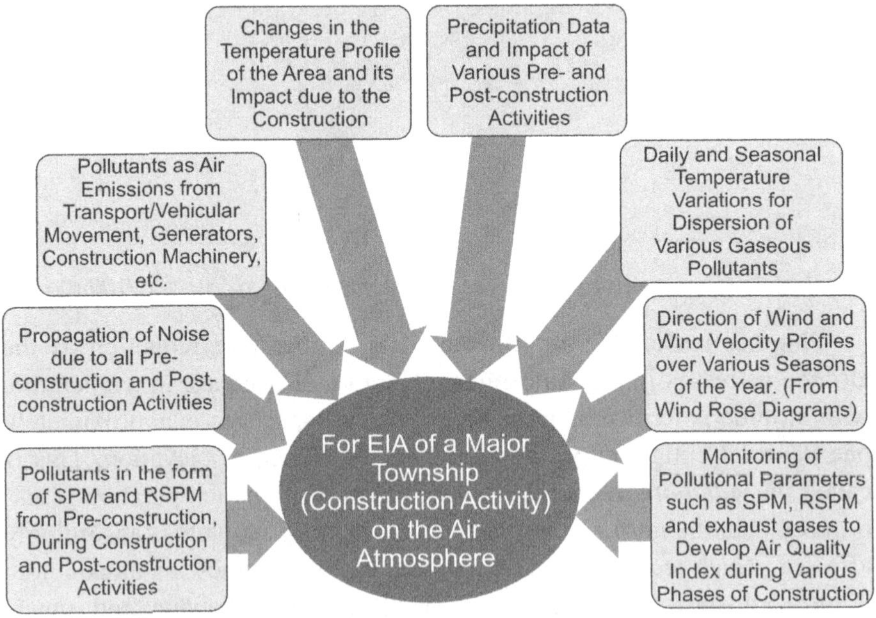

Fig. 5.5 *Environment impact assessment, meteorological data and potential impacts on air atmosphere*

The famous air pollution episode of London smog of 1956 was because the pollutants were trapped in the lower atmosphere due to a colder air layer on top, which acted as a lid on the warm pollutant gases such as oxides of sulphur below and did not allow the gases to disperse in the atmosphere. The concentration of the pollutants in the trapped layer increased and caused a large number of deaths in the public.

5.3 INDUSTRIAL STACK EMISSIONS AND CONTROL STRATEGIES

When continuous release of air pollutants takes place from a stack, it is referred to as a "plume" while with intermittent release it is called a "puff". The shapes and patterns of plumes are shown in the Fig. 5.6. They are caused due to wind, temperature and weather (sunny or overcast) conditions, etc. Dispersion in the atmosphere depends upon the height at which the pollutants are released, the exit volume and velocity and temperature of pollutants and the atmospheric conditions prevailing at that time.

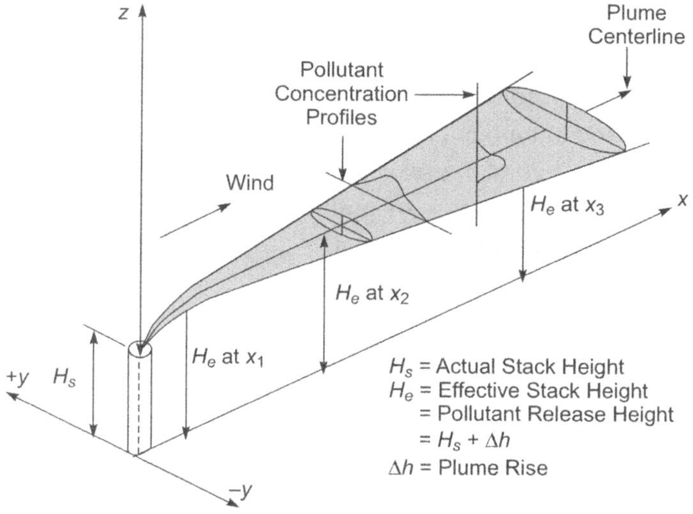

Fig. 5.6 *Plume dispersion along three axes.*
(Ref.: http://en.wikipedia.org/wiki/Air_pollution)

The pollutants coming out of the stack get dispersed in all the three directions, namely, the longitudinal direction along the direction of wind, lateral to the wind direction and in the vertical plane, i.e., along Z-axis. Figure 5.6 explains the reduction in concentration along the three directions as the pollutant moves away from the stack. Downstream air quality is affected by the location, topography of the terrain and atmospheric conditions at the time of release. The system is complex and needs mathematical modelling to get answers.

5.3.1 Control Strategies

The following typical control strategies are generally adopted by industries for abating air pollution from their factories and meeting their national standards:

1. Avoid pollutant formation
   This is best done by using cleaner fuels such as those with low sulphur content, low ash content, no lead in petrol, etc.
2. Undertake process change such as the following:
   - Use Titanium dioxide instead of lead paint pigments
   - Steel shot blasting instead of sand blasting
   - Wet methods instead of dry methods
3. Localise spread of pollutant by having separate enclosure / layout for certain industrial areas
   - Having separate Foundries, Paint shops, Plating shop from the rest of the workshop.
   - Enclose conveyors to prevent dust/chemicals flying in the inside or ambient atmosphere.
   - General good housekeeping.
4. Reduce concentration of pollutants before disposal by adopting necessary control equipments
   - Gravity settling
   - Cyclones (dry or wet)
   - Bag filtration
   - Electrostatic precipitation
5. Dispose by dilution in atmosphere

This is the final and most important step. The height of the stack depends on the nature of emissions, the terrain, the wind velocity, temperature inversion conditions, etc. As several variables are involved and climatic conditions also vary from season to season, mathematical modelling may be required to find the minimum stack height and other features so that pollutant concentrations at the ground level at the desired site downwind are within acceptable limits prescribed by Control Authorities. As a thumb rule, the higher the stack, the lesser the ground level concentration obtained.

A full-scale Environmental impact study may also be required to cover other aspects such as water, wastewater, solid wastes, etc., besides air pollution (see Chapters 5–8). The control authorities may prescribe various restrictions and even require that certain industrial operations be shut down during certain adverse atmospheric conditions.

As an example, the accidental release of toxic chemical, methyl isocyanate (MIC) at Union Carbide, Bhopal, can be quoted. The season was winter which aggravated the effect on the windward side while there was very little or no effect on the leeward side of the release. The Bhopal gas tragedy is termed as one of the worst chemical disasters on earth. It killed about 15,000 people and injured about half a million. Genetic effects have also begun to show up in the next generation.

5.4 TRAFFIC POLLUTION AND ITS CONTROL

Urban areas globally have large number of vehicles, which is a major source of air pollution in such areas. Emissions from vehicles are emitted at ground level and remain for varying periods in the atmosphere adjoining the ground levels depending upon the atmospheric conditions. The emissions are at more or less the breathing level for majority of the urban population.

The basic chemical reaction taking place in an automobile engine is combustion of fuel, i.e., combustion of hydrocarbons. If the chemical reaction is ideal and complete, it results in carbon dioxide and water. If not, it will result in CO_2, water and other gaseous pollutants depending on the type and quality of fuel, conditions in the engine cylinder or mainly efficiency of engine. Air pollutants arising out of traffic emissions are:

Hydrocarbons

Carbon Monoxide (CO)

Nitrogen Oxides (NO_x),

Sulphur dioxide,

Particulates (respirable size $< PM_{2.5}$ and others $\geq PM_{10}$,) volatile organic compounds (VOCs) and smog.

Prolonged and continuous exposure to the air pollutants is responsible for increased incidences of respiratory ailments at large. The sensitive population such as children and aged group of urban dwellers is more affected because of comparatively lesser tolerances against the effects of air pollutants.

One study found higher concentration of Lead (Pb) in samples of teeth collected from the kids residing by the side of a metropolitan city's main road which had heavy vehicular traffic. Lead is also found to go piggyback on dust particles in the air. Lead is a known to be a toxic heavy metal which affects the kidney and brain (lowered intellectual ability) and other human systems. Now, "Lead" is no more used as an additive in automobile fuels (unleaded petrol). The major causes of traffic pollution are shown in Fig. 5.7.

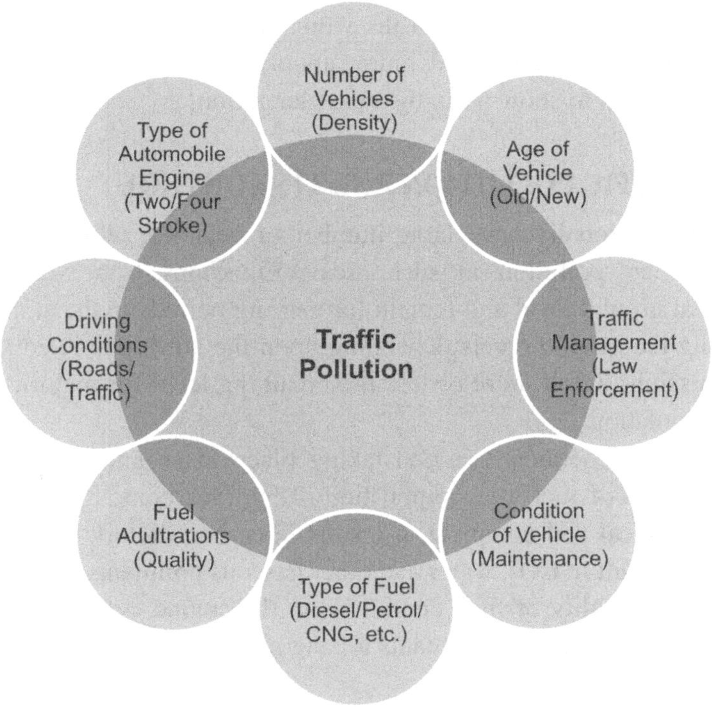

Fig. 5.7 *Major causes of traffic pollution*

Traffic pollution is an important major problem concerning health of the city dwellers. Traffic pollution can be controlled by applying control measures separately or in combination depending upon the type of traffic, infrastructure, technology available and its affordability, awareness and participation of the people, legal framework and political will of the region or area.

Control of traffic pollution is possible by adopting some of the following measures or their combinations:

1. Technological measures:
   (a) Upgradation/improving efficiency of the automobile engines. Reducing emissions has had concurrent benefit of ensuring better fuel efficiency.
   (b) Making feasible the technology using alternate energy sources for automobile engines, i.e., Hybrid/electric cars, etc.
   (c) Making present fuels more environmental friendly such as better quality fuels (avoiding addition of lead to petrol to improve its octane number).
   (d) Developing exhaust treatment technologies such as catalytic converters.
2. Traffic planning and management:
   (a) Easy availability of mass transit systems.
   (b) Improving the driving conditions (Road conditions).
   (c) Elimination of traffic bottlenecks/traffic jams (Standstill traffic).
   (d) Discouraging and restricting entry of vehicles, cars, etc.
3. Legal/fiscal policy issues related to vehicular pollution:
   (a) Stringent air pollution norms (for monitoring authorities).
   (b) Implementation of pollution norms to manufacturers and users.
   (c) Restricting entry depending upon type and age of vehicle.
   (d) Road tax and vehicle tax benefits to environmentally alert users.
   (e) Heavy penalty/Environment tax to violators.

Present scenario of the traffic pollution is that majority of the controls as suggested above are already on its path of implementation such as unleaded petrol, availability of better quality fuels, CNG engine kits, better and efficient automobiles, construction of flyovers and road bye pass/ring roads, automated traffic signals, car pooling, environment cess/tax, etc. Stated measures will have a positive impact on the total quantity of air pollutants entering in urban atmosphere.

We must remember the fact that the number of vehicles are increasing at an alarming rate. Participation of individual users, urban dwellers, automobile manufacturers, policy makers and policy implementers, all will be needed to help in controlling traffic pollution significantly. The public is becoming more and more conscious of the need to control traffic pollution. Support for use of blended fuels, hybrid cars, and electric vehicles is increasing day by day. Also, larger cities are understanding the advantages of having more and more mass transport systems from their business point of view (see Chapter 2 and Section 11.1).

5.5 INDOOR AIR POLLUTION AND VENTILATION

Normally, houses in warm climates have to keep their doors and windows open to remain comfortable. In colder climates, however, people have to keep their doors and windows closed for the same reason. When doors and windows are open, the air quality inside the house is more or less the same as the air outside which already comes under the purview of the local authority for ensuring its quality.

Indoor air pollution and ventilation become important when doors and windows are kept closed for any of the following reasons:

- ➢ Climatic reasons (domestic houses, offices, etc., may keep their doors and windows closed only in cold weather or all the year round to provide air conditioning).
- ➢ Social reasons (Auditoriums and cinema halls wish to provide air conditioning for comfort as well as social reasons).

The study of indoor environments from the air pollution perspective has become important because of urban life styles. The pollutants present in the inside atmosphere may be any one or all of the following: chemicals, insecticides, adhesives, paints, volatile organic carbons (VOCs), natural pollutants such as radon, from the soil or structural materials, smoke from wood and other fuels used for cooking, heating, etc. in the house.

In air conditioned environments, volatile organic chemicals, along with other air pollutants tend to remain in the inside atmosphere for a long time affecting the inmates. Finer particles (RSPM) tend to get recycled in air conditioned indoors and build up in concentration while the larger sizes get filtered out in recycle.

The cooking places or "chullahs" (used for cooking where wood or coal is burnt as fuel) and fire places tend to pollute the indoors. Smokeless chullahs can help in such climates to protect the inmates from the harmful effects of air pollutants.

A Scientific study has also expressed concern about the indoor atmosphere providing environments conducive to certain micro-organisms. Permanently air conditioned atmospheres (e.g., ducts) if not regularly cleaned, are likely to promote growth of harmful microorganisms which may affect the inmates.

Some common pollutants found in the indoor atmospheres are:

1. Particulates from dust from Indian stoves (chullah) using coal, wood, dung, etc.
2. Asbestos from roofing and other items.
3. Carbon monoxide from stoves, cigarette smoking, etc.
4. Harmful volatile organic chemicals from furniture, paints, solvents.
5. Nitrogen oxides from gas stoves and kerosene stoves.
6. Sulphur dioxide from burning of coal and kerosene.
7. Radon from soil.

The indoor atmospheres where industrial activities are carried out (workshops) can have variety of other air pollutants than those listed above. These depend upon the kind of activity carried out in the enclosed workshop. The air changes needed for an enclosure to ensure good ventilation are said to be 30 per hour. Nonetheless, the norms for safety of the inmates given under the Factories Act have to be met.

As per the National Building Code the area of openings for ventilation and lighting purposes should be minimum 15% of the inside area. In case of unavailability of minimum opening, the ventilation and lighting is to be provided by artificial means.

Natural and proper ventilation is important as it can improve the quality of air in the indoor atmospheres by:

- Reducing the concentration of indoor pollutants.
- Improving the thermal comfort conditions in indoor atmospheres.
- Decreasing energy consumption needed in air conditioning.

It should be remembered that the extent of natural ventilation that can take place in a room depends on the scope there is for cross-ventilation. No ventilation will take place in a room where only one window is open. It needs a minimum of two windows (one for inlet of air and one for its outlet) to get natural ventilation.

Cross-Ventilation

The extent of air that can come into a room depends on the extent of air that can exit from the room. Hence the exit often becomes the controlling feature.

Secondly, some gases are heavier than others and stratification may occur. Thus generally, outside air must enter from the lowest level and exit from the highest level otherwise the heavier gases (like Chlorine) will settle to the bottom (floor level) and tend to accumulate there.

The following two examples illustrate how a workshop can become polluted with a chemical and how its concentration can be estimated.

Problem 5.3: If 10 grams of coal containing 5% sulphur are burnt in an enclosed room of 30 m^3 volume. Estimate the sulphur dioxide concentration at 1 atmospheric pressure and 25°C. Assume all sulphur is converted to SO_2

Sulphur burnt = 0.05 × 10 grams

= 0.5 gram and SO_2 formed = 1.0 gram

In terms of weight/volume, we get

$$1000 \text{ mg } SO_2/30 \text{ m}^3 = 33.33 \text{ mg/m}^3$$

Problem 5.4: How many grams of Chlorine gas must leak out from a chlorination equipment to give a concentration of 100 ppm in an enclosed room of 80 m^3 at 1 atmosphere pressure and 25°C temperature.

$$\text{ppm} \times \text{mol wt} / 24.436 = \text{mg/m}^3$$

(Molecular weight of Chlorine is 35.5)

$$100 \text{ ppm} \times 71 / 24.436 = 290 \text{ mg/m}^3 = 0.290 \text{ g/m}^3$$

In a room having 80 m^3 volume, the leak has to be 0.29 × 80 = 23.2 grams.

5.6 OZONE DEPLETION AND ACID RAIN

5.6.1 Ozone Depletion

Ozone is one gas which saves life on earth from the scorching effect of solar radiation. The Ultra Violate (UV) rays from the Sun are trapped in the Ozone layer present in the upper atmosphere. The stratospheric Ozone is at an altitude of approximately 30 km, and the thickness of the Ozone layer is just a few millimetres.

This Ozone layer acts as a shield to stop the entry of UV rays of certain specific wavelength which have proved to have a carcinogenic effect on human skin. The people, especially with white skin (deficient in the colour pigment "Melanin" of human skin) are more likely to be affected leading to what is called as Melonoma (cancer of the skin). The usage of sunscreen lotions capable of protecting the skin from harmful effects of UV rays is common in certain geographical locations of the globe.

The natural shield of ozone against UV radiation was observed to be punctured at the poles. The reason was found to be usage of certain chemicals such as Chlorofluorocarbons (CFCs). The CFCs are inert and not soluble in water, and this property kept them in the atmosphere without changing their chemical composition. When these chemicals reached the Ozone layer high up, they entered into a chain reaction with ozone converting the ozone molecule to oxygen. Once the reaction is triggered it continued creating an "ozone hole" by converting ozone to oxygen. Hence this void or gap in the ozone layer allowed harmful UV radiation to penetrate into the atmosphere causing harm to people on earth.

Realizing the disastrous effects, as a part of global measures it was decided to ban the use of CFCs as refrigerants and replaced them with alternate chemicals. Global elimination of ozone layer harming chemicals, have shown results and the scientists have confirmed that the ozone layer in the atmosphere is being slowly reformed to its natural state.

The ozone, which is formed as a result of photochemical smog (NO_x, VOCs, UV and particulates) due to the automobile exhaust during the daylight, is harmful. It affects our respiratory system. The ozone has oxidising properties and has harmful effects on those exposed to it. This ozone is termed as ground level ozone. In some parts of the world, the local authorities alert their citizens to the presence of such pollutants in the urban atmosphere.

5.6.2 Acid Rain

Fossil fuels (coal and oil) carry impurities in the form sulphur. Coal is used after only physical sizing and it is not possible to remove the sulphur from it. Though the refining process of crude oil is aimed at removing the impurities, it is not feasible to remove all the impurities. When these fossil fuels are burnt say in the automobile engines or power plants, along with the hydrocarbons the sulphur also oxidizes to oxides of sulphur.

The combustion process uses air for burning and the nitrogen also gets oxidized to oxides of nitrogen. These gases reach in the upper atmosphere where the formation of sulphurous acid and sulphuric acid take place. Similarly the nitrous and nitric acids are formed from oxides of nitrogen reacting with the atmospheric moisture. The carbon dioxide from the atmosphere dissolves in the moisture to give carbonic acid. All these chemical reactions are complex and take place under specific conditions in the atmosphere. The cocktail of these acids in the aerosol form gathers in the atmosphere and comes down with precipitation (rain). This is termed as "acid rain".

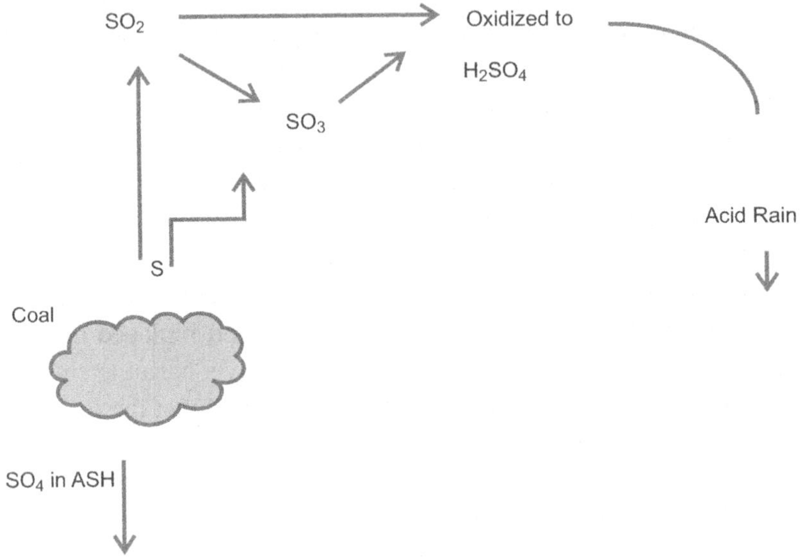

Fig. 5.8 *Acid rain*

Effect of acid rain on the soil is that it has the potential of dissolving the micro-nutrients present in various forms in the soil, turning fertile soil into barren soil. The surfaces of metals, concrete, stones, marble are corroded and lose their economic value.

The example of world famous Taj Mahal of Agra, India, where the white marble started turning yellow with other chemical changes in the marble (calcium carbonate) due to the neighbouring industries at Firozabad and Mathura. The corrective steps were taken to control the emissions and now the effect on the Taj Mahal is controlled. Gaseous emissions from the various manmade activities are responsible for the acid rains. Minimising use of fossil fuels by switching over to renewable energy is one way.

5.7 NOISE POLLUTION

Noise pollution may not be considered so serious as the other forms, but affects the well-being of a person and affects the quality of the environment. Noise is to some extent a personal matter. One may dislike any form of sound when what the person wants is just plain quiet.

Noise is measured in decibles (dB) and since noise is transient, not accumulative, it is best noted in the form of a survey of an area. A library-like noise level is about 40 dB, light traffic registers as 75 dB. The ambient noise levels permitted in certain zones depends on the limits prescribed by the local authority such as

Area	Prescribed Range, dB
Residential	45–55
Commercial	55–65
Industrial	65–70

(It may be noted that 1 dB increase means double the intensity.)

Control Measures

Sound is best controlled by use of appropriate strategy either in an industry or an area or a street or near a hospital or school building. For example, one may use of:

- ➢ Sound absorbing materials in industries.
- ➢ Sound absorbing mounts and pads.
- ➢ Spring mounts.
- ➢ Isolation of noisy machines in separate rooms.
- ➢ Process change.
- ➢ Legal/punitive measures (for traffic noise, festival (Divali) noise, etc.).

Usually, elaborate rules and regulations exist, but the usual problem is lack of enforcement by control agencies.

> **Observation:** As far as air pollution control is concerned the following biblical phrase seems most appropriate to use:
>
> "DO UNTO OTHERS AS YOU WOULD HAVE THEM DO UNTO YOU"!

EXERCISE QUESTIONS

1. In measuring particulates why is size ≥ 10 μ differentiated from size ≤ 2.5μ? Give examples.
2. In traffic pollution, which are the parameters of concern, and why?

3. Distinguish between episodic and chronic noise pollution? What limits would you prescribe in each case?
4. Explain the causes of "ozone depletion" and "global warming".
5. Assume you are planning to set up a new workshop to manufacture some chemicals which are odorous and have an explosive characteristic when a certain concentration is reached. How will you ensure natural ventilation in the workshop to avoid such an eventuality? Also, what steps will you take in locating the workshop.

PROJECT/ASSIGNMENT TOPICS

Briefly discuss each of the following options for implementation, marking those which you think are more feasible to adopt, and those which might possibly have undesirable side-effects.

Soft Options

- Monitor new projects to see if all approval conditions are implemented.
- Appoint more monitoring staff.
- Levy more stiff fines if implementation is lacking.
- Hold seminars and training programs.

Hard Options

- Implement a program to train the trainers for stack and other monitoring work.
- Promote consultancy services to advise industries on air pollution work.
- Adopt cleaner technologies in manufacture. Build up a roster of examples.
- Develop standards for specific air pollutants and their measuring methods.

6 CHAPTER

Water Resources and Supply Systems

6.1 Principles of Water Treatment and Supply

6.2 Drinking Water Quality for General Consumption

6.3 Water Quality for Industries and Boilers

6.4 Development of Water Resources—The Indian Scenario

6.5 Conclusion

Remember the Chinese proverb of Chapter 2: "Drink water, Think source".

6.1 PRINCIPLES OF WATER TREATMENT AND SUPPLY

Every town in every country is able to grow and sustain itself because it has a water supply system serving it. Without a water supply system it would not be able to develop beyond the village level. Villages become towns and towns grow into large cities only when piped water supply, and other infrastructure, are provided.

6.1.1 Water Sources

The source of fresh water may be any one of the following:

- A flowing river
- An impounded or a natural lake
- A ground water source

In each case a system of water supply has to develop to serve a town. Commerce begins to thrive. More industries start to come in and life goes on as we know it.

6.1.2 Water for Different Uses

Water supply norms for municipal areas vary from country to country depending upon their water resources, their climate, their consumptive needs and lifestyle of the people. In India, the supplies to cities and metropolitan areas stand today (2012) between a minimum of 70–90 L/person/day to as much as 200 L/person/day. Below 70 L/person/day the sewers would be difficult to keep clean while beyond 200 L/person/day (assuming that water resources permit the rate) it would be considered too lavish a use of water by the Indian public.

Table 6.1 gives the break-up of water for different uses in urban households. Table 3.1 given earlier in Chapter 3 gave the break up where rainwater harvesting was done implying careful, minimal usage. Table 6.1 gives the range of values typically observed in Indian urban towns.

Table 6.1 *Approximate water usage for different purposes in towns and cities*

	Use of Water	Estimated Daily per Capita Requirement (L/Cap-day)
1.	Only Drinking	2–5
2.	Drinking, cooking and minor uses	2–20
3.	Bathrooms, Washing, Cleaning, etc.	40–90
4.	Toilet flushing	40–90
	Approx. Total	90–200
Gardening requirements can be estimated on the minimum basis of 7–10 L/m^2–day But this water may not be of drinking water quality.		

For industries the water requirements can vary quite substantially from industry to industry. The following few values shown in a tabular form have been taken from various sources for a few industries. Actual values may vary from country to country and plant to plant.

Cane sugar	500–1000 L/tonne
Meat Slaughter houses	5 m^3/tonne live weight of slaughtered animals
Fruit and veg canning	8–80 m^3/tonne
Soft drinks	2–5 L/L of product
Alcoholic beverages	10–20 L/L of product
Pulp and Paper	190–230 m^3/tonne
Soap manufacture	>200 m^3/tonne

6.1.3 Water Quality and Quantity Fluctuations

Quality and quantity is what everybody is concerned with the raw or untreated water fluctuates both in quality and quantity, the extent of fluctuation depending on its source and the season of the year. The monsoon season is generally the worst for water quality because rain and run-off lead to soil erosion and wash down of various natural substances as well as applied chemicals such as pesticides, fertilizers, etc. Flow in rivers is also maximum during the rainy season.

River Waters

River water fluctuates most in both quality and quantity, seasonally. Besides this, it is a patent ecosystem fact that rivers are a source of water for the people as well as a depository of the wastewater after it is used by them. (See Chapter 1)

Rivers may be subject to floods when the wash-down of silt and debris is maximum. At other times, the flow may be more steady but the town or city's wastewater discharged into the river may affect its quality and quantity. As explained in Chapter 1, organic wastes may undergo stabilization in the river as it flows, while inorganic substances may only accumulate in concentration within the river. Such waters may need varying degree of treatment depending on the season. Fish prefer well oxygenated areas within a river. They will swim away from areas where the dissolved oxygen is low.

Concentration of accumulative pollutants increases after each city or town.

As Fig. 6.1 shows, the oxygen profile dips after leaving a city as organic matter from sewage exerts oxygen demand. Minimum oxygen situation occurs between two adjacent towns. Fish swim away from such areas, or die in them if oxygen is too low.

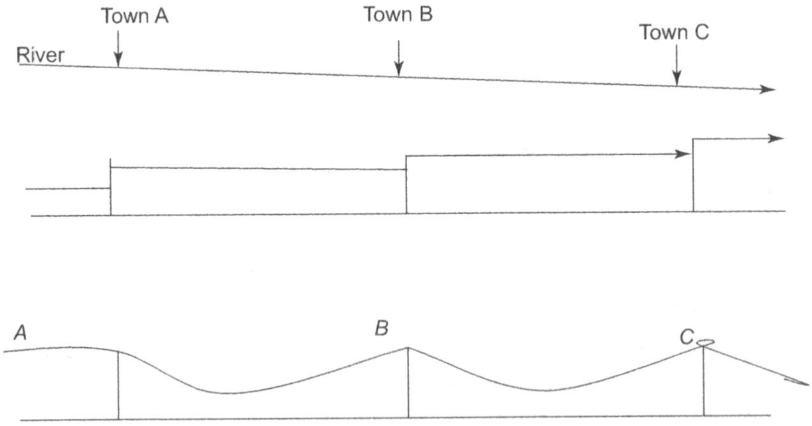

Fig. 6.1 *Fate of degradable and non-degradable pollutants entering a river from three cities A, B and C located on it*

Lake Waters

Raw lake waters generally show lesser fluctuation in quality and quantity as they undergo natural settlement and blending in a lake environment. Lake waters, however, may show the effects of algal and aquatic vegetation growths, especially at certain times of the year. Lake waters also reflect the nature of activities going on along the banks of the lake and its catchment area surrounding it. There is an expression "like father, like son!" to show that the lake waters reflect whatever is happening on land around the lake (Fig 6.2). So, lake water quality control begins with the nature of development permitted in the catchment area. If the catchment area is clean, the lake has a good chance it will also remain clean.

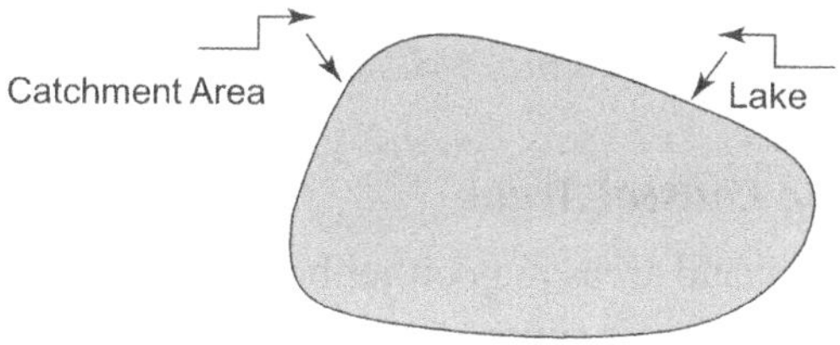

Fig. 6.2 *Lake quality depends on type of activity going on in the surrounding catchment area*

> One can stretch the well-known genetic analogy "Like father, like son!" to apply to a lake because the lake's water quality reflects the land-use going on in its surrounding areas. An ecosystem phenomenon exists as explained in Chapter 1.

Groundwaters

The waters are extracted from the ground by sinking one or more tube wells as shown in Fig. 6.3. The tube wells are operated by motors on top which drive submersible pumps below the water line. At the lower end of the tube well is located a foot valve. Regarding ground water quality, the soil may have already provided filtration and the water quality may be perfect for meeting drinking water standards without any further treatment. However, in some cases, dissolved impurities may be present and treatment may be necessary for their removal.

Ground waters generally show the least fluctuations, quantity-wise, with the seasons. However, in the dry season groundwater levels may fall so low that for all practical purposes no supply is possible. The reader is referred to Section 3.2 to understand that employing proper rainwater harvesting methods can overcome this problem by replenishing the ground waters and raising their levels so that water is available even in the later part of the dry season.

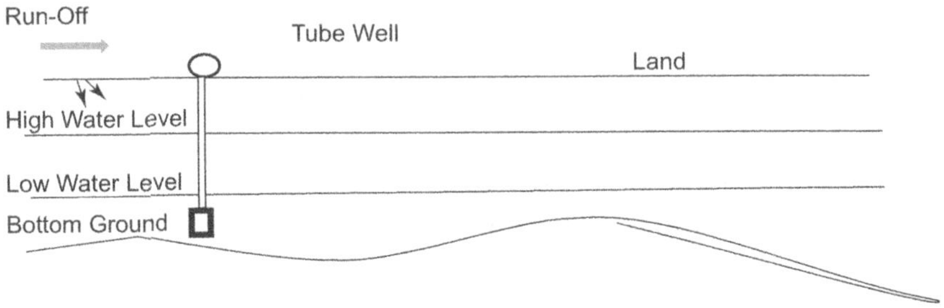

Fig. 6.3 *The figure shows a tube well arrangement to extract groundwater*

Groundwaters in Coastal Areas

Groundwaters in coastal areas may either be stratified into two layers (salty below and sweet above) or mixed up to give what are called "brackish" waters.

Often, a clear separation occurs based on the specific gravity of the two waters. Sea water is heavier than fresh water. Sea water ingress occurs inwards into the land along the coastline to some extent, while at the same time sweet (fresh) water from rainfall percolates into the soil and flows seaward. Where the two waters meet, a stratification tends to occur because fresh water is lighter and tends to float on the heavier sea water. This is why people living in coastal areas often successfully sink *shallow* wells to tap the fresh water. If they draw too much fresh water, it invites the sea water to come in and take its place, thus giving brackish water.

6.2 DRINKING WATER QUALITY FOR GENERAL CONSUMPTION

Water is known as the best solvent in the world. Therefore, it stands to reason that there will be a large number of substances present in it, of which some may be harmful to health, some imparting only a taste or colour to the water, and some just present in the water. Water analysts have grouped them in general as follows.

General parameters

Flow

Suspended sediment

Oxygen

pH

Total dissolved solids (TDS) or Conductivity

Temperature

Black listed parameters

Arsenic (As)

Mercury (Hg)

Lead (Pb)

Cadmium (Cd)

Halogenated organic compounds (e.g., DDT)

PCB

Grey list parameters

Ammonium ion (NH_4^+)

Nitrite ion (NO_2^-)

Nitrate ion (NO_3^-)

Total N

Total P

Chromium (Cr^{+6})

Nickle (Ni)

Zinc (Zn)

Organophos compounds

Phenols

BOD

Coliforms

Water quality for general consumption has to meet minimum standards prescribed by the Government (which is generally based on the international recommendations of the World Health Organization). The standards specified by the Government are legally enforceable.

The recommendations as approved by the Ministry of Health, Government of India, are available to all water authorities in India.

6.2.1 Water Treatment

Whatever the source of water, some treatment is necessary before it is supplied. The main principle behind water treatment is to make water drinkable without affecting health. We separate out the coarsest material first by plain/gravity settlement of the raw water, followed by chemical dosing to coagulate several smaller particles into larger particles which can then be removed by gravity. The next step in treatment is to provide sand filter to filter off still smaller particles, and finally we kill the rest of the very fine particles (the microbes) by using a disinfectant. Dissolved impurities are then removed. In other words, an assembly line technique is followed, so to speak, removing from the largest to the smallest size – the smallest being micro-organisms which are killed in the end.

The sequence of steps in water treatment can be shown symbolically as in Fig. 6.4:

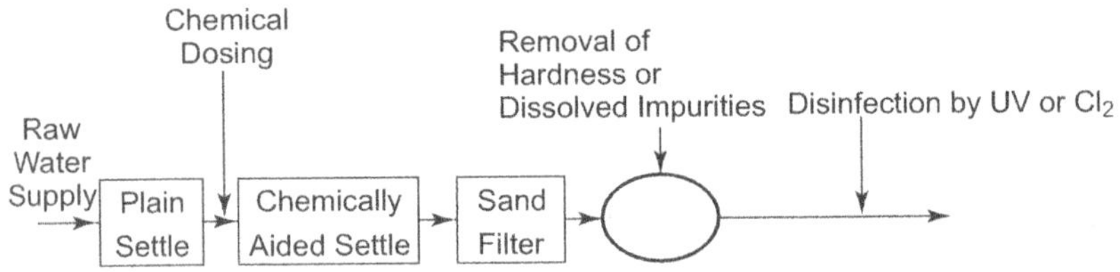

Fig. 6.4 *A typical water treatment flowsheet*

All the above steps may not always be needed. For example, if it is a lake water, the first plain settling step may not be needed. If it is a ground water, nature itself may have already removed all the coarse and fine particles, and no filtration may be needed. Only dissolved impurities (such as hardness, or iron and manganese, or other metallic compounds or fluorides or pesticides) may need to be removed. This may be done using ion-exchange methods or membrane filtration. The final step of disinfection is always provided to ensure safety in case of public supplies. Even when dealing with best quality spring water where no other treatment is needed, disinfection is done.

After treatment, the water may be stored in a sump or reservoir for use as needed. Care is taken to ensure that no insects or mosquitoes can breed in them as otherwise malaria or dengue mosquitoes can breed. Supply is made either by gravity or by pumping (depending on the topography) through a distribution system of pipes laid below the roads all through the town to serve the people.

Cost of treatment depends on the quality of the raw water and the finished quality desired. Various other factors, some site-specific (like pumping) and some industry-specific also affect the cost. Hence, no attempt is made to give any cost figures here.

6.2.2 Quality and Quantity Checks in Distribution Systems

After the treated water leaves the treatment plant and enters the piping system, its quality and quantity can both be affected negatively. Two things have to be generally assured in water distribution systems serving towns and cities: Quality and quantity. Leakages can reduce the pressure, and leakages can also be a source for entry of polluted groundwater into the pipe during non-supply hours.

1. Water quantity has to be supplied everywhere at a required minimum pressure so that the water can rise at least up to the 1st or 2nd floor height of buildings to be served in the town. Higher floors can be served by additional pumping by the building occupants.
2. Water quality has to be fit for drinking. Quality has to be checked for its physical, chemical and bacteriological parameters before the water enters the distribution system. Once it enters the distribution system its quality can still deteriorate in regard to its bacteriological quality due to pollution seeping in. For this purpose, a minimum strength of disinfectant (usually residual chlorine) is maintained everywhere in the distribution system at say a prescribed value (e.g., 0.20 ppm) at all times. Absence of residual chlorine would signify bacterial/organic activity which would render the water quality as "doubtful". In this manner, both quality and quantity are controlled by the water supply authority.

Water samples are drawn from different parts of the distribution system from time to time and tested for its bacteriological, chemical and physical quality using standard laboratory testing methods detailed by the Government in consultation with the World Health Organization (WHO). For physical and chemical parameters, several specific tests are prescribed depending upon the source and known quality of water and have to be carried out accordingly. For bacteriological quality, the water sample is tested for an indicator organism instead of several different organisms. This indicator organism is *Eschericia coli* (commonly referred to as *E. coli* or coliforms). Coliforms by themselves are not pathogenic (disease-causing) but as their source is the same as that of excreta (namely, the intestinal tract), they are used as an indicator organism. Their presence is presumed to indicate likely sewage pollution and need for caution.

The sad story of Mumbai's deteriorating water quality should be mentioned here. Roughly 25% of their daily water samples FAIL in their bacteriological tests. This is mainly due to neglectful maintenance and ageing (old) leaking water mains, neither repaired nor replaced as excavation is difficult owing to heavy traffic, narrow streets and dense population demanding water. During non-supply hours, the leaking pipes leak in the reverse direction, so to speak, and sewage polluted water from the ground enters the pipes and mixes with the drinking water supplied to the public. A failure of 25% is indeed high.

What a Distressing Situation!

Roughly 25% of daily water samples in Mumbai FAIL in their bacteriological tests owing to leaking, old, water mains and poor maintenance.

6.2.3 Bottled Water

The same high quality standards which apply to drinking water also apply to bottled water, perhaps even more so! The sources of bottled water are generally four:

1. Pure mountain water which is generally melted snow or glacier,
2. Clear Spring water, and
3. River or ground water, or
4. Municipal water demineralized and re-mineralized to strict standards.

The source of water is what gives the water its special brand value. Bottled water has to be very carefully treated, bottled and marketed because they usually become "branded" products in course of time. Their brand value brings them good price. Their physical, chemical and bacteriological quality has to always meet the high standards expected of them.

6.3 WATER QUALITY FOR INDUSTRIES AND BOILERS

Several manufacturing processes are normally quite satisfied with the quality of water which meets drinking water standards. However, certain industries, or certain industrial processes have more stringent quality requirements and their quality has thus to meet standards prescribed by the Indian Standards Institute (ISI) and in some international cases by the International Standards Organization (ISO). These standards are obtainable from the standards institutions concerned or the various industry associations involved.

Boiler water quality generally depends on the pressure at which the boiler operates. The higher the pressure, the better the quality has to be. However, the quantity required may be relatively small compared to the quantity required by the whole factory. Hence, only a small fraction of the total requirement may need to be given additional treatment to suit the situation.

Similarly, water of relatively poorer quality (e.g., reused water with minimal pre-treatment) can often be provided for toilet flushing, washing vehicles, washing shop-floors and roads, cooling waters, transporting materials like logs of wood and wood chips, carrying out washing operations where water quality is not important, etc. Waters meeting high quality can also be obtained from reused water, where necessary. Thus, different qualities of water may be involved in case of industries, such as the following:

- Waters meeting drinking water standards (potable waters)
- Waters meeting boiler feed-water and higher standards, and
- Waters of non-potable standards.

This gives us a tremendous scope for reuse of water, as we see later in Chapter 8.

6.4 DEVELOPMENT OF WATER RESOURCES—THE INDIAN SCENARIO

Water resources development in India has traditionally meant building more and more new dams, reservoirs, canals, tube-wells, etc., to continue increasing fresh water supplies to meet ever-growing urban, industrial and agricultural needs. This is what one may call the "*upstream*" side of water resources management. It consists of developing either new water resources or making better use of existing fresh water supplies through water conservation. We have naturally concentrated on this upstream side as long as new fresh water supplies were available in a region. In some regions, however, rainfall has been scanty; in some regions plentiful; in some regions the soil has been silty-sandy; in some clayey. Only recently, are we trying various methods of rainwater harvesting to improve our resources.

Lately, we have also started to think of the "*downstream*" side of water resources management, namely, pollution control and reuse of wastewater after treatment so as to use the water that has been used once. Just as necessity is the mother of invention, so necessity has led us to adopt reuse and recycling of used water after treatment in order to augment our water resources and use the available water more efficiently than before.

In spite of taking many upstream measures water resources have been proving unsustainable in many regions and much emphasis is now being laid on *downstream* measures such as pollution control, recycle and reuse of wastewater not only for non-potable purposes in industry and agriculture but even for augmenting regular community supplies for potable purposes.

> The upstream side of water resources management includes: all fresh water supply sources, fresh water conservation measures and rainwater harvesting. The downstream side of water resources management includes all used water, used water pollution control measures, and reuse and recycling of used water after treatment.
>
> Wastewater itself has now become a water resource and a more holistic approach is beginning to emerge, integrating both upstream and downstream measures so to achieve a good "mix" of water resources development and management activities.

6.4.1 Use of India's Water Resources

As stated earlier, India's water resources are reported to be used, roughly, as follows:

Agriculture	70–80%
Industry	10–15%
Municipal including drinking	10–15%

As is common knowledge, more water resources are needed in every area as populations increase, industries increase and lifestyle changes take place. In some Indian rivers today, however, the fresh water flow is only 1/2 to 1/3 of what it was just 30 years ago. Just 30 years ago, Mumbai could supply its people 150 L/person/day. Today the city can hardly supply 90 L/person/day. What will be the situation after another 30 years? Besides reduced supply, the quality is also getting bad. 20% of Mumbai's samples fail daily to meet bacteriological standards! Drinking such water without proper treatment can give disease.

Global warming is adding to our problems. It is making our fresh water resources erratic from year to year. Sometimes too much, sometimes too little. In coastal areas, sea level rise brings the sea water more inland and makes ground waters brackish in those areas and unfit for both agriculture and human consumption. So our whole water resources picture is going awry.

6.4.2 Water Conservation

Water conservation is the first thing that comes to mind when there is shortfall in water supply. The reader is referred to Section 1.4 in which various water conservation measures are described in detail. Water conservation measures are generally required to be taken in every city whether shortfall in water supply occurs or not. Why should any water be wasted at all?

The next step generally taken by water authorities to meet any shortfall situation is to encourage rainwater harvesting and reuse of water after treatment for some specific uses.

6.4.3 Rainwater Harvesting—A Boon or a Mixed Blessing

In cities like Mumbai, rainwater harvesting is "recommended" by the municipal authorities to meet water shortages, but in doing so, it must be remembered that it divides responsibility between the municipality and the people for water quality. When quality goes wrong in an area each party tends to blame the other and nothing is often done to prevent a repeat. Besides quality, malaria and dengue are on the increase in cities like Mumbai. Rainwater harvesting on individual basis in a city leads to several sites dispersed all over the city where mosquito breeding can occur. Rainwater harvesting in a city is actually a mixed blessing!

6.4.4 Rainwater Harvesting is Good for Rural Areas

Rainwater harvesting is good for rural areas here public water supplies have not yet been fully established. In fact, better water management in rural areas can change the face of India. Many stories are told about *Anna Hazare* who changed the face of a village called Ralegan Siddhi (some 80 km from Pune). He brought about social upliftment by introducing various measures but mainly through rainwater harvesting to ensure increased agriculture through groundwater storage. It also improved horticulture and the growth of (mango) trees. The result was a prosperous village which everyone now wants to copy.

The author visited another village in Vansda district in Gujarat, to see the good work done by another social worker *Dr Kirtibhai Vaidya* (a medical doctor who gave up his lucrative practice in Mumbai and devoted his life to social upliftment work in Gujarat). Both these "gurus" have shown that better water management increases availability of groundwater for agriculture and makes rural farmers shift from a struggling, poor one-crop economy to a more prosperous multi-crop economy. This is what India needs. Better water management can change India's economy, and also slow down migration to cities. Water has a tremendous role to play in our lives.

As we explained in Chapter 3, Rainwater harvesting is more suitable, in fact essential, in rural areas of India to ensure that more groundwater recharge takes place which can later be used as and when required for agricultural and drinking purposes. In rural areas, rainwater harvesting can be a boon.

6.4.5 What about Cities and Towns who Suffer Shortages Year after Year?

They suffer mainly because of any one or all of the following reasons:

1. The price at which water is sold is not right. Water is sold so cheap that water conservation is neglected. In fact, in some States of India, free electricity is given to farmers for pumping up groundwater, with the result, some never stop the pumps even at night! Admittedly, people are poor in India. So the price of water has to be kept low. What costs us ₹ 12–18 per 1,000 liters to produce, we sell for ₹ 3. We try to make up our losses from industry by cross-subsidy. The problem is that low price makes water conservation unimportant for the people in general.
2. Enough funds are not generated for timely expansions. We have to depend on politicians in our country and in the World Bank and elsewhere for funds.
3. Expansions are not conceived early enough. Much time is lost in planning and taking permissions. The system is unable to keep up with rapidly increasing demand.
4. Poor operation and maintenance, leakage and wastage takes place.

Additionally, there can be a water quality problem. The raw water itself is deteriorating in quality these days through the presence of pesticides and various metals and heavy metals washed down in the water after agricultural use. Their removal only means upgrading the water treatment process, which cannot be done in the absence of funds. Comfortable 24-hour water supplies have slipped into intermittent supplies with its concomitant difficulties, and also giving the public a cocktail of unwanted chemicals. Fire-fighting has been crippled for want of water.

> Water supply problems have never deterred urban growth, just inflicted suffering.

Perhaps three things can possibly be done (besides water conservation) where there is a shortage of water year after year. In fact, for large cities like Mumbai all three of them may have to be done at the same time to benefit from their relative costs and features:

1. Rainwater harvesting
2. Reuse of wastewater after treatment
3. Desalination

Rainwater harvesting We have already pointed out the potential health hazards of rainwater harvesting in urban areas when done on an individual basis. Nonetheless, cities like Mumbai are already implementing rainwater harvesting programs on individual building basis to supplement city supplies. This is not recommended in the larger interest of public health, but has to be tolerated as a temporary way of meeting shortages. Mumbai's municipal lakes are anyway doing rainwater harvesting on a large scale and they are followed by city-wide supply under full municipal responsibility for quality. This is acceptable. But, we should not divide responsibility between municipality and the general public who has to do rainwater harvesting on an individual basis or buy tanker water of unknown quality to meet shortages in supply.

Reuse of water – the Indian or American way

> **The Indian or American Way?**
>
> Reuse of wastewater after treatment is feasible to do in cities in two ways: either the Indian way on an individual building basis or the American way on a city-wide basis through two separate networks laid in the city, as explained in the text.

In the US, reuse is being done on a city-wide basis where a duplicate network of pipes is laid and dual quality water (potable and non-potable) is supplied through two different networks. In California (57 MLd) and Florida (85 MLd) are supplied in this manner.

In Indian cities, laying an additional set of pipes at this stage to carry dual quality supply does not seem feasible as it would cause enormous traffic problems for long periods. Our population densities are also high and roads are relatively narrow. Moreover, the people's widespread lack of awareness of health effects of waters of different qualities would also cause problems. Remember the severe jaundice epidemic that Delhi had in 1955 when it was inadvertently supplying treated and untreated Yamuna river water to the city. Since that time in India, it is preferred to treat wastewater and reuse it on an individual building or individual industry-wise basis where water quality can be better controlled.

Reuse of wastewater on individual building basis During Mumbai's water shortage days in early 60s, reuse was developed first for the Air-India building in Mumbai in 1963 when the building's sewage was collected in the basement of the building, treated and reused to make-up cooling water losses in air-conditioning. There was inherent safety in the arrangement because the wastewater was used after treatment in an entirely different circuit (for cooling purposes).

Thereafter, treatment for reuse was done again in Mumbai for a large industry in Chembur, on an individual basis (for Union Carbide, now Oswald Agro). They picked up wastewater from an adjoining municipal sewer and reused it after treatment in their various industrial operations. Several plants of this type came up subsequently. Thus India can boast of over 40 years experience with treatment for reuse (also see Chapter 8).

Reuse of wastewater on a citywide basis There was another Indian way of doing reuse on a citywide basis which was different from the American way: In Hyderabad, in the early 90s, it was suggested that wastewater could be treated and added to the lake for city-wide reuse.

Indian lakes are different from western lakes. We lose 30–50% of lake volume in evaporation alone owing to our hot weather. Spreading chemicals on the water surface to control evaporation has not been very successful in India. Well-treated wastewater can be used to top-up a lake and make up for evaporation losses. This was first implemented at Hussain Sagar Lake in Hyderabad in 1998 by use of reclaimed wastewater (20 MLd) to make up for water lost in evaporation. It is still being done. It has been possible to do this in the Hussein Sagar Lake because water from this lake is not used for potable purposes.

> **The Hyderabad Experience**
>
> The question often asked is: Can reuse be done as in case of Hussein Sagar Lake, Hyderabad, if the lake was indeed, used for potable purposes? The answer is "yes", if a natural treatment step is provided for the wastewater after regular tertiary treatment.

Whatever the extent of tertiary treatment given (primary, secondary and tertiary) to the wastewater, it would be wise to have it followed by a "natural" treatment step in the form of groundwater recharge. In flowing underground, the wastewater's "identity" would be lost and further "natural" treatment would be received. Nature can do what we cannot always do. After groundwater recharge, water can then be introduced into the lake without fear of any public health problem.

Indirect reuse of sewage is being done internationally in USA, Europe and Israel. Singapore is beginning with reuse, using membrane filtration since land is not available.

Desalination This is the final form of reuse. It is obviously limited to coastal areas where the sea is near at hand. Besides location, desalination has two major problems: Firstly, it is expensive as it needs much electric power to operate it (₹ 6–7 per L) comparable to the cost of bottled water). And, secondly, while generating electric power by conventional means (oil or coal), it co-produces CO_2 which is not favoured in today's global warming environment. The latter problem can be avoided if a renewable source of energy is used specifically for the desalination plant.

6.4.6 Suggested Linking of Rivers in India

As stated in Section 1.4, a novel way has been suggested in India of ensuring that rivers in the south of India be fed with surplus waters from the relatively water-rich rivers of North India, by linking the northern rivers with the southern rivers. This idea has been very actively favoured by many authorities dealing with water resources in India. However, one wonders if this will truly end up as a panacea for water shortages in Southern India, or cause some unexpected problems?

Perhaps from a quantity viewpoint, this may prove to be a great (though expensive) idea. However, from a quality and ecosystem viewpoint this may prove a disaster which only time will tell. The waters of the North are generally snow-fed and therefore cold. Their introduction into the warmer waters of the Southern rivers will perhaps cause a kind of thermal pollution in reverse and affect the microorganisms unfavourably. The whole ecosystem will no doubt eventually adjust to the new temperature and food regime with several species (perhaps some of commercial value) may be lost in the transition. This is what we do not know. All we know is that there will be an effect and the effect will cause a ripple of other effects to follow. The case of the US experiment of deforestation and how it affected water quality over a wide area (cited earlier in Section 1.1) may also be recalled. A similar situation will arise in India. Who knows whether linking of rivers in India will after all prove to be a good idea or an ecological disaster waiting to happen! When the disaster happens, it will be too late to retrieve our steps. Conservation and reuse seem to be a better idea to begin with.

> The million dollar question is how far are we prepared to go with the three alternatives given above, namely, rainwater harvesting, reuse/recycle and desalination, for a city that is short of water?

6.4.7 How far are We Prepared to Go?

This is a three-pronged question, a "Psycho-Socio-Economic" question! If money was not a problem, a city like Mumbai would go for desalination Plants like some of the Middle-East countries (but, hopefully, with a renewable source of energy, not oil). However, money being a problem, the answer would be *all* available methods would be used in a "mix" that we can afford for the city in question.

Naturally, we would begin by adopting water conservation and harvesting rainwater for augmenting water supply (in spite of its health hazard potential) as that would probably be cheaper to begin with. At the same time, reusing wastewater for irrigation instead of fresh water is something all cities and towns must examine as, topography permitting, it would be the best way to conserve fresh water for drinking and using wastewater instead for irrigation. Further, the wastewater when travelling under-ground would in fact improve in quality through "natural" processes and be ready for reuse either by villages further downstream or for lake replenishment, depending on topography and soil.

Reusing wastewater after tertiary treatment for "non-potable" uses (toilet flushing, washing cars, gardening, industrial uses, etc.) in individual buildings as well as in industries as done in Mumbai, would have to continue being done in the future also. It is something we will all have to practice more efficiently in the future.

Feeding treated wastewater to a lake for potable purposes to make up for evaporation losses (after tertiary treatment and additional natural / land treatment, as well as with good monitoring and control measures in place) is another thing we will have to learn to do well in the future.

Finally, we will have to have desalination plants to add to our water resources in coastal areas like Mumbai, Chennai, etc. (operated with an independent renewable source of energy).

Beware!
Water supply is a monopoly business in which there is no competition. This can make any municipal body slow, lethargic and unresponsive to the situation. Pricing of water can also get out of sync.

We are Running Out of Water Resources

In fact, we are running out of water resources and if we are not to perish for want of water, we must seriously find new sources of water as soon as we can and we must develop water conservation and reuse of water as best as we can. A typical sequence is:

Develop fresh water sources → Promote water conservation and reuse → Add more fresh water sources (if possible) → Promote more reuse (as possible) → Finally adopt desalination

There is no escape from desalination someday for us to meet our ever-increasing water requirements because of our large and ever-increasing population. The danger is we may then have a local power crunch unless we generate enough power from renewable sources for desalination, and reuse wastewater without damaging the environment.

6.4.8 Working Experience

Coming to think of it, we already have working experience with most of the above items. We have already made a start in Hyderabad and Rajasthan with advanced forms of reuse.

Maharashtra and Gujarat are reported to have made progress with groundwater recharge at many sites. Narkhed, 60 km from Nagpur, meets its water shortage by this method. Maharashtra's effort to keep Chandrapur Power Station running at full capacity by reusing the city's wastewater upto 32 MLd as cooling water after treatment, is remarkable. It will keep a public utility working in spite of water shortage and will also provide much operating experience.

However, we have no working experience as yet with some items. Feeding treated wastewater to make up for evaporation losses in a lake which is used for potable purposes would be an interesting exercise for India to undertake at a suitable site.

Mumbai's lakes like Tansa, Vaitarna, Modak Sagar, etc., might also be considered someday for this purpose as land for groundwater recharge might be available at sites nearby, but their problem at present would perhaps be non-availability of enough sewage or other wastewater in sufficient quantity nearby to treat and feed into the lakes. This must be left for further investigation. The MCGM (and most other Corporations) have the requisite ability for placing the required good monitoring and control measures in place.

Similarly, experience now needs to be gained in operating a desalination plant in India. In this matter, it would be ideal if the temptation can be resisted to draw the necessary electric power from an existing grid. In deference to global warming considerations, a stand-alone renewable energy source (perhaps solar or wave energy) needs to be set up along with the desalination plant. Thus, valuable working experience would be gained with operating both desalination plants and renewable energy systems for the future.

6.5 CONCLUSION

It is clear that we are responding, though perhaps a bit too slowly, to the need for a holistic approach in planning our water resources, integrating both upstream and downstream measures. We are also beginning to find that water and wastewater are two sides of the same coin and wastewater is actually a dependable water resource in some cases that we can no longer neglect.

We are also learning that water availability, especially for agriculture in rural areas, is a determinant in our country's prosperity. In that sense, India's future depends on its water supply and how it is managed! Better water management can change India's economy, and slow down migration to cities.

REFERENCES

Guidance Manual for Drinking Water Quality Monitoring and Assessment (2007) CSIR-NEERI, Nagpur and NICD, New Delhi, Supported by WHO.

Recommended Additional Reading

It is suggested that the reader refers to the internet to find related items of his interest for additional reading.

EXERCISE QUESTIONS

1. What is the status of drinking water resources in India? What immediate steps do you think can be taken?
2. Why do you think water conservation should always be the first step?

PROJECT/ASSIGNMENT TOPICS

1. Assume you have been invited by the Central Railways (CR) to identify ways of minimizing water consumption at the Chatrapati Shivaji Terminus Mumbai (CSTM). Following data are available

   On an average 25 lakh commuters pass through the station daily. There are 1,000 guest rooms provided by CR for accommodating the passengers and the train crews that operate long distance trains. These trains when they arrive at CSTM, get cleaned, washed, repaired, after which engine attaches at other end for return journey out of Mumbai. 300+ such trains are managed in and out of the station. In addition, nearly 1,000 local trains touch CSTM every day. Nearly 50% of the local trains are also cleaned, washed and repaired in the CSTM workshop. Needless to say the dozens of eating joints, tea stalls, mess facilities serve the 1,000 rooms and administrative offices of CR, 365 days a year. Where would you like to take measurements:

- To estimate water consumption at the CSTM
- To estimate quantity of wastewater generated at the station
- And suggest a brief plan for water audit

2. Briefly discuss each of the following options for implementation, marking those which you think are more feasible to adopt, and those which might have undesirable side-effects.

Soft Options

- Request people to conserve use of water in industries and agriculture.
- Hold seminars and discussions on increasing water resources and local water supplies.
- Adopt better management information systems.

Hard Options

- Prepare a long-term and a short-term plan for developing water resources.
- Implement a program to train the trainers for leak detection work.
- Protect water quality by protecting surrounding land use.
- Switch over from intermittent system to a 24 × 7 system, and reuse where possible.
- Secure Governmental/International Funds for system expansion.
- Increase the price of water to make water supply profitable for future expansion.
- Plan measures to protect habitation in coastal areas against sea level increases.

7 CHAPTER

Wastewater Treatment for Pollution Control

7.1 Some Concepts in Wastewater Management

7.2 Objectives of Treatment

7.3 Wastewater Treatment Technologies, Factors in Choice

7.4 Resource Recovery, Waste-to-Energy

7.5 Some Treatment Methods for Domestic Wastewater

7.6 Will Total Sanitation Ever Come in Our Lifetime?

7.7 Industrial Waste Treatment and "Zero" Discharge from Industries

7.8 Thermal Pollution

7.9 Marine Pollution, Coastal Waters (CRZ), Mangroves and Wetlands

7.10 Coastal Waters and Coastal Regulation Zone (CRZ)

7.11 Common Effluent Treatment Plants (CETPs)

> *Yes, we can get any pollutant out of water (at a cost); but when "politics" gets into water there is precious little that we can do about it!*

7.1 SOME CONCEPTS IN WASTEWATER MANAGEMENT

7.1.1 Wastewater

"Wastewater" can be a very complex substance varying from domestic to a wide array of industrial wastewaters. Domestic wastewater includes typical domestic (human) sewage, excreta or feces, urine, sputum, soap, bathroom wastes and kitchen wastes including grease, grit, etc. Wastewater from toilets is often called "black" waters, while wastewater from bathrooms, kitchens, etc., are called "grey" waters.

Industrial wastewaters include a large variety of wastes from different manufacturing industries mixed in different proportions. Municipal wastewater is usually a mixture of domestic and industrial wastewater in different proportions, and is generally described in very broad terms as a mixture of organic and inorganic matter in dissolved and suspended form, containing grit, grease, various chemicals, minerals, nutrients and micro-organisms. Municipal wastewater is often like domestic wastewater in characteristics.

BOD/COD

Wastewater being a complex substance, its "strength" is generally indicated by its *demand for oxygen* at a known temperature (e.g., 20° C) and can be expressed in two ways, either as "Biochemical Oxygen Demand" (BOD) or as "Chemical Oxygen Demand" (COD). In both cases, the higher the oxygen demand, the greater is the presence of organic matter.

If the BOD test is carried out over 5 days, it is called BOD_5 test. If it is carried to its ultimate value (which generally takes more than 5 days) it is called BODu test. The ultimate value is naturally higher than the 5-day value.

BOD with separate and combined sewers In developing countries like India located in seasonal rainfall (monsoon) areas *separate* sewers are generally preferred, one for carrying the rainwater run-off during the monsoon season and another for the daily domestic and industrial wastewater in the non-monsoon season. In countries like India where rainfall occurs in the monsoon season only, separate sewers are used. The BOD_5 of domestic wastewater in such cases is generally of the order of 45–54 g/person/day at 20° C. In countries like the UK where it can rain at any time throughout the year (e.g., most of Europe), *combined* sewers are preferred and the 20° C BOD_5 of combined wastewater is around 40% higher (77g/person/day).

Industrial wastes have widely varying BOD_5 depending on the industry, its raw materials used and the manufacturing processes involved.

7.1.2 Daily Wastewater Production in Urban Areas

The type of sewerage system to be provided depends on the availability of water for sanitary purposes, as shown below:

Daily Water Usage (L/person/day)	Sewer System Provided
Less than 25	Dry latrines, boreholes, eco-san, etc
25–75	Localized drainage. Septic tanks
70 or more	Public sewerage system possible with self-cleansing velocity (centralized or decentralized)

Daily water usage in India varies from less than 25 L/person/day for unsewered rural areas to as much as 90, 120, 150 L/person/day or even more in urban areas. Towns with a minimum flow of about 70 L/person/day can be considered for providing a sewerage system since experience shows that a minimum flow of about 70 L/person/day appears to be necessary to ensure flow through sewers to keep them flowing and not allow the excreta to be left stranded in the sewer lines. At this flow, the sewer can be designed to have a "self-cleaning velocity". A higher flow of fresh water may be allowed in some cities where water resources permit it.

A rule of thumb says at least 80% of the water supplied to a community returns as wastewater. The rest is likely to be consumed in cooking, drinking or lost in leakage and evaporation. Thus, when Mumbai was supplied water @ 180 L/person/day, the wastewater was probably = 0.8 × 180 = 145 L/person/day and the BOD_5 was probably = 54 g/per/person × 1000/180 = 300 mg/L. Now that the water supply rate has halved, the BOD must have doubled in value to nearly 600 mg/L.

If the rate of water supply of a city is low, the BOD_5 of the wastewater is proportionately higher. Chennai has a water shortage and, therefore, has an even higher BOD_5 value of the sewage than Mumbai.

For domestic wastewater, if the BOD_5 is around 300 mg/L, the BODu value may be around 1.5 times the BOD_5, i.e., around 450 mg/L. The COD value is about 1.6 to 1.9 × BOD_5 value, namely COD may be around 480–570 mg/L. For industrial wastewaters the values can vary substantially. For an industrial waste rich in organic matter (like a distillery waste) the BOD_5 may be in the thousands.

Breakdown of Organic Matter

As stated earlier in Section 1.1, nature has so provided that organic matter is unstable and degrades into inorganic end-products. This happens *in any case* whether oxygen is present or not, because there are three sets of "decomposers" present in nature: aerobic, facultative and anaerobic, to tackle any situation.

Aerobic decomposers operate in the presence of free oxygen. If free oxygen is not available, facultative organisms take over and obtain their oxygen from breakdown of nitrates (NO_3). After nitrates are exhausted, the sulphate molecule is broken down to obtain oxygen. When sulphates undergo breakdown sulphides are formed. As everyone knows, sulphides give the smell of rotten eggs and is a sure sign that anaerobic activity has started. Finally, only anaerobes take over till anaerobic breakdown is completed. The nutrients released from the original organic matter are not lost, but recycled to support the photosynthesis process once again provided water and sunshine continue to remain available.

7.2 OBJECTIVES OF TREATMENT

7.2.1 Objectives

Treatment of wastewater generally has one major and all-important objective to protect the general environment and public health. This objective is sometimes split into its two components, health and environment, although in reality we cannot have one without the other.

Environmental protection generally implies stabilizing the organic matter and releasing the nutrients for recycle to keep the ecosystem going, while health protection implies destruction of pathogens and preventing transmission of persistent and accumulative pollutants through the food chain. Generally, both objectives are met by meeting the effluent discharge standards and drinking water quality standards of the country.

7.2.2 Total Sanitation and its Status in India

According to the World Health Organization, however, health protection requires total sanitation. Total sanitation is essential for obtaining tangible health benefits and includes "sanitation" of all forms of water and wastes in a community such as:

- Drinking water
- Wastewater (black water + grey water)
- Storm water
- Municipal solid waste (garbage)

In any country, sanitation is difficult to achieve on a countrywide basis. It becomes doubly difficult task when by sanitation we mean "total" sanitation. In India, even our major metro-cities are yet having a less than ideal water supply system and are only partially covered by a sewerage system. We have yet to install total sanitation systems in most of our cities and towns. Our country is so advanced in many ways, and yet so backward in some others. Let's hope we can kickstart the subject of sanitation, soon.

> It is sometimes joked that we have 5-star hotels in "zero-star" cities!
>
> 'I think water and sanitation are not getting the attention they deserve because they are not sexy! On the contrary, satellites, computers, software and industrial growth are sexy!'
>
> Dr Sam Pitroda

In a large and democratic country like India, continuous awareness generation and advocacy are essential tasks when we have to deal with over a billion people. Most of our NGOs are already conversant with this subject. Hence, we will not dwell on it here. Besides awareness, appropriate technologies have to be used. Several organizations have been actively developing more affordable, more acceptable and more easily manageable technologies with which our poorly staffed local bodies can maintain the existing infrastructure.

Additional problems complicating sanitation in India are: our rapid growth rate, poor water supply situation in many parts, general need for water conservation, the difficult electric power position, the difficult land availability position, and severe lack of funds.

> The question we all have to ask is how can we speed up sanitation coverage under the circumstances?

7.3 WASTEWATER TREATMENT TECHNOLOGIES, FACTORS IN CHOICE

7.3.1 Wastewater (Sanitation) Coverage

Sanitation includes at least three stages: the defecation stage, the conveyance stage and the treatment and disposal stage. To get the utmost value out of the usually small budget, one has to use the most appropriate technology for the stage and area in question, and ensure people's participation through education and training to stop open-field defecation. Hopefully, the following guidelines will help:

On-site

Where water is scarce one may not be able to have a water-carried sewerage system and may have to be satisfied with one of the following "on-site" dry disposal systems (see Fig.7.1)

- ➢ Bore-hole latrine
- ➢ Double-pit latrine
- ➢ Eco-san

> **The best example one can give of an ordinary person's environmental concern is the case of our own Mahatma Gandhiji.**
>
> He is reputed to always carry a small spade with him wherever he went. He knew he would have to defecate in the open fields as there would be no toilets where he stayed in villages. He used the spade to dig up some earth to cover up the excreta instead of leaving it in the open for flies to sit on. This is exactly what we now recommend to do in case of on-site disposal and in case of "sanitary landfill" of solid wastes.

On-site septic tanks Another on-site disposal system often used is the septic tank which is preferred when water is available for using flushing toilets but no sewerage system exists to carry away the wastewaters. A septic tank followed by soakage in a pit or trench is one of the most commonly used methods all over the world. Often readymade septic tanks are available fabricated in different sizes, and the required size has to be ordered. They can also be easily fabricated at site from brick and mortar.

Septic tank size is generally fixed such as to give approximately 24 hours detention time for the wastewater (both grey and black wastewater together) in a rectangular shaped tank. Thus *capacity = no. of persons × water supply rate per person per day*. The hydraulics of the tank can be improved by providing baffles at the inlet and the outlet. The efficiency of treatment expected in passage through a septic tank is about 30% BOD removal (20–50%). Hence the effluent from a septic tank is always disposed by soakage into the ground through soak pits or trenches. Hence, soil has to be porous enough for gradual absorption to occur.

Percolation rates generally vary from 50–100 L/m^2/day) depending on type of soil. Sometimes, to keep the size of installation small, only black waters are passed through a septic tank while the grey waters are used directly in gardening or agriculture.

Off-site

Where enough water is available for providing a sewerage system (>70 L/cap/day), the sewage may be collected from an area and made to flow to a site where some treatment can be given before discharging the effluent to an adjoining water body land or land for irrigation. This is called "off-site" disposal because the waste is generated at one site and disposed of at another site. Some treatment is usually required to be given before final disposal.

7.3.2 Sewer Standards and Discharge Standards

Sewer standards and discharge standards are two different standards. Sewer standards are concerned with treatment which may need to be given before the waste enters the sewer line. This may be necessary to ensure sewer line protection. *Sewer standards* are usually laid down by the local municipality either for ease of maintenance of its sewer system by their workmen or for protecting the sewers from destruction by corrosion or explosion or chokage.

Discharge standards pertain to final disposal of treated sewage to a water course or an irrigation field. Discharge standards may require an additional treatment step to meet the final disposal requirements (laid down by the National Pollution Control Authority) for discharge of the treated effluent either to land for irrigation or to a water body (river) as the case may be.

7.3.3 The Sewerage System

Two types of sewerage systems are usually provided: (1) A conventional centralized sewerage system in which there is a network of sewer pipes laid as a main central line along with branches and laterals and having a major outfall at the end, or (2) A decentralized sewerage system in short lengths serving smaller areas and having several outfalls. The sewerage system adopted is important as in a way it often fixes the treatment method to be selected (natural or mechanical as explained below).

In the case of a conventional centralized sewerage system, a mechanized method of treatment has to be generally used (like Activated Sludge or Trickling Filters) because the flow is relatively large and has to be discharged to the nearest water course (river) after treatment. A decentralized sewerage system is generally provided to cover a smaller area. The final effluent volume is therefore relatively smaller and a "natural" method of treatment can be used followed by groundwater recharge. This is beneficial as it helps to conserve water instead of discharging to a river which ultimately goes to the sea. Thus the sewer pattern can affect both treatment method (natural or mechanized) and water conservation practice.

Natural Methods of Treatment

These are those which depend on the action of "natural" plants or organisms like algae, duckweeds and micro-organisms all of which draw their oxygen from natural sources (not mechanical aeration, nor moving machinery):

- Bore-hole and pit latrines
- Septic tanks
- Oxidation ponds (algal ponds)
- Duckweed ponds
- Constructed wetlands
- Land irrigation systems

Natural methods have no machinery in them, but need more land to give the desired performance. They are cheaper and more reliable to operate as they need no power or machinery for operation. Natural methods are most desirable for small and medium-sized towns of India where land is available but skilled operation is difficult to find.

Figure 7.1 shows a simple natural treatment method using an Oxidation Pond in which algae flourish. The oxygen produced helps to support aerobic micro-organisms which stabilize the organic matter contained in the sewage.

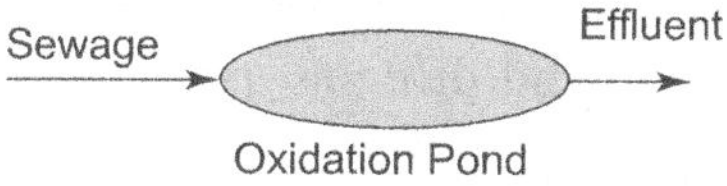

Fig. 7.1

Simple Oxidation Ponds, for example, work because of the action of micro-organisms and algae. Oxidation Ponds get their oxygen from two sources:

1. Direct exchange of oxygen from the air above the pond at the air-water interface, and
2. The oxygen released along with algal growth in the pond (see Eq. 1 in Section 1.1).

Aerobic micro-organisms act as decomposers in the pond. If enough oxygen is not available, facultative micro-organisms take over and breakdown of nitrates starts in order to secure oxygen from them. Thereafter, breakdown of sulphates begins with anaerobic organisms. When this happens smelly sulphides are formed, and sewage treatment plants with the smell of rotten eggs are hated by all.

Overloaded wastewater treatment plants give typical hydrogen sulphide smell. One way of overcoming the smell problem is to either increase the size of the pond so that oxygen supply is increased, OR one may artificially add a nitrate compound to the pond so that oxygen supply is increased through breakdown of nitrates – in which case the sulphates do not need to be touched and smell nuisance is avoided. The decision depends on relative costs and availability of land.

The performance of natural systems is somewhat dependent on the seasons (winter/summer) as temperatures and other natural conditions like sunlight vary with the seasons. This makes Pollution Control authorities relatively hesitant as they prefer equally good performance throughout the year.

Mechanized Methods of Treatment

These are often used (especially with larger flows) as their performance can be more or less even throughout the year. Also, mechanical systems need lesser land which makes them particularly suitable where land availability is a problem in treating large flows. This happens especially near larger cities and metropolitan areas. However, mechanized methods need electric power for mechanical aeration and moving machinery makes operation more expensive and difficult to maintain. A few examples of mechanized systems are the following:

- Activated Sludge/Extended Aeration
- Sequential Batch Reactors
- Trickling Filters
- Aerated Lagoons
- Rotating Bio-discs
- Anaerobic Sludge Digesters
- Membrane Filtration, etc.

With mechanized methods (see Fig. 7.2) treatment generally begins with screening of sewage and plain settling to remove as much organic matter as possible before expensive aeration is done. The settled sludge is sent to anaerobic sludge digesters for treatment before drying.

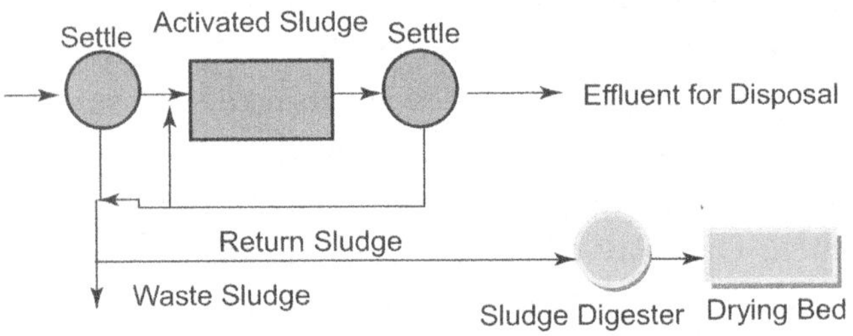

Fig. 7.2 *A mechanical treatment plant using the activated sludge process to stabilize organic matter contained in the sewage*

Choice between Natural and Mechanized Technologies

Before a choice can be made about the treatment technology to be used, the "local constraints" must be identified such as

- Availability of land
- Availability of reliable, 24-hour power supply
- Availability of skilled maintenance staff
- Final effluent quality desired to meet discharge standards

Land is the most important requirement of all the above items. Where land is available, all "natural" systems of treatment are generally possible, and perhaps more appropriate to adopt in case of domestic sewage and municipal wastewater.

In choosing between the various natural and mechanized systems, one has to keep in mind the three keywords noted below. The system selected must be:

1. *Affordable,* to the people who pay for it
2. *Acceptable*, to the people and authorities who approve of it, and
3. *Manageable* for the people who will operate it

On all these three counts, natural systems would easily win. They are generally the most "affordable" because expensive electric power is not involved in operating them. Also, where no electric power is involved there is no concurrent production of greenhouse gases. For the same reason namely, absence of mechanization, they become most "manageable" for the relatively less skilled staff available in small and medium-sized towns. Normally, natural systems are also "acceptable" to the neighboring people and the governmental authorities who have to approve their installation. It is only on rare occasion, especially at higher latitudes that their performance in colder winter months may not be up to the mark.

All mechanized systems require electric power to function and thus are relatively more expensive in operating costs and produce greenhouse gases. They are therefore, less affordable. For the same reason (mechanization) they become less "manageable" for the relatively poorly skilled staff available to municipal bodies. However, for industries this may not be much of a problem as they may already be having a large pool of skilled staff for operating their factory equipment. Normally, mechanized systems are also "acceptable" to the neighboring people and the governmental authorities who have to approve their installation. Hence, a mechanized treatment plant may work out to be quite "appropriate" for an industry or a large city. The only question may remain which specific process is most suitable for the wastewater under consideration. This depends on several local factors and economic considerations which are beyond the scope of this book.

7.4 RESOURCE RECOVERY, WASTE-TO-ENERGY

7.4.1 Resource Recovery

Resource recovery is strictly speaking not an essential objective of treatment. It is an incidental advantage if some resource can be recovered and money earned so as to reduce the net cost of treatment and to make it more meaningful and more sustainable. Resources commonly recoverable from wastewater treatment are:

1. Fuels such as bio-gas.
2. Soil conditioners/fertilizers such as sludge.
3. Foods such as fish, fish-food (such as algae), geese, ducks, irrigated crops.
4. Reusable water.

Every effort is normally made to recover as much resources as are possible depending on the type of wastewater treatment method adopted – natural or mechanized. Recovery of reusable water is, however, becoming increasingly more important as fresh water shortage is becoming more and more pronounced.

The only guideline that can be given at this stage is that all waste (either solid or liquid) which contains organic matter can be– should be– converted into useful "energy" or food. In today's world, every source of energy must be used for whatever best use it can be put to. Thus, the phrase "waste-to-energy" plants is an important phrase. Examples of waste-to-energy plants are:

1. Gobar gas plants for excreta disposal in rural areas.
2. Upflow Anaerobic Sludge Blanket (UASB) plants for wastewater treatment.
3. Other biomethanation plants for solid wastes treatment (also see Chapter 9).

> What we want are more "waste-to-energy" plants, not waste of energy plants!
>
> We want plants capable of recovering resources, rather than using our scarce resources (like land and electric power).

7.5 SOME TREATMENT METHODS FOR DOMESTIC WASTEWATER

Table 7.1 gives a few recommended methods for treating domestic and municipal wastewaters in different situations.

Table 7.1 *A few treatment methods suitable for different situations*

Type of Area	Treatment System	Effluent	Disposal	Examples
Unsewered, rural, water-scarce area (<25 L/cap-day)	Natural	Only ablution water	On-site, Underground	(1) Aqua-privy, (2) Double-pit (3) Eco-san, etc.
Unsewered area, Suburban or rural, Water < 70 L/ cap-day	Natural	Anaerobic	On-site, Underground	Septic tank + soak pit
Sewered area, urban or rural, Water @ > 70 L/ cap-day	Natural	Aerobic	Off-site, Surface	(1) Septic tank + Constructed wetlands (2) Oxid Pond + Irrigation. (3) Duckweed+ Fish+ Irrigation (4) Aer Lagoon + Oxid pond + irrigation
Sewered metropol area, Water @ > 70 L/cap-day	Mechanized	Aerobic	Off-site Surface discharge Or reuse in industry	Conventional treatment or "waste-to-energy" plant plus addl treatment before reuse

Generally, in most of India, one of the following four combinations of treatment methods can be recommended:

1. ***For individual or small groups of bungalows with on-site disposal***
   Septic tanks +Soak pits
2. ***For districts with several houses and off-site disposal***
   Septic tank + Constructed wetland + Ground water recharge
3. ***For small and medium-sized towns***
   Oxidation ponds + Land irrigation + Ground water recharge
4. ***For larger cities, metropolitan areas***
   Aerated lagoons + Reuse in local industry

A few examples of the range of land and power requirements of the different treatment steps are given below in a tabular form:

Treatment	Land	Power
Activated sludge	about 0.25 m^2/person	2.0–2.5 HP/1000 persons
Aerated lagoon	0.3–0.4 m^2/person	2.0–2.5 HP/1000 persons
Oxidation pond	1.0–2.0 m^2/person	Nil
Constructed wetland	1.0–2.0 m^2/person	Nil
Irrigation field	100–200 m^3 wastewater/day/ha	Nil

Aerated lagoons do require somewhat more land than activated sludge; but aerated lagoons are far simpler to operate and maintain. The aerators can be fabricated in standard sizes in conveniently located workshops and the earthwork for the lagoons can give welcome employment to the local people and use local materials.

For cities sometimes instead of aerated lagoons, the UASB is recommended as it is anaerobic in nature and does not require electric power. However, being anaerobic, corrosion can be a severe problem making it difficult to maintain. This also requires use of special non-corrosive materials and special concrete work which increases the capital cost. Indians prefer aerated lagoons which are easier to maintain and do not require too much more land.

Among all the mechanized treatment methods, facultative aerated lagoons are the simplest to operate and their land requirement is quite low but they have not been used as much as they should because their power requirement is quite considerable and a continuous source of power is necessary. Facultative aerated lagoons are therefore an excellent candidate for use of non-conventional energy sources such as wind and solar systems.

Natural or Mechanized?

Natural systems of treatment are simple, non-mechanized systems which are reliable, cost-effective and easy to maintain. They are especially suited to our warm climate. Yet, we often run after Activated Sludge and other complicated systems which are more appropriate for colder climates! In colder climates they cannot use natural systems.

When the simplest mechanical system (e.g., the aerated lagoon) is used in combination with a natural system like the oxidation pond, the designer can optimize for land or power or annual cost, and make a very economic and simple design – best suited for India's towns and cities of different sizes. Think of it!

7.6 WILL TOTAL SANITATION EVER COME IN OUR LIFETIME?

Will any of these sanitation measures ever be taken in our lifetime? In a democracy where peoples' level of awareness is at a relatively low level, as in India, things can take long to happen. When you have to carry everyone with you, "awareness generation" and "advocacy" are important routes to follow, however slow they may be. And a beginning can be made most profitably by starting with young people of school age. Since almost half of India at present is said to be below the age of 25 years, awareness generation would seem ideal to begin with young people.

Awareness generation does not imply just talks, lectures and posters. The actual condition of the school/college water supply and its toilets, the quality of home water supply and the state of home toilets are all important to improve as they leave a lasting impression on the young mind. Habits are formed from then on. So schools and local bodies also have to do their bit.

Another group that keeps a long lasting impression of the situation is "tourists". Tourists are not only foreigners, but also the nationals of our own country. Every young and old person in a country is a potential tourist in his own country. All people notice the sanitary facilities that exist outside their own home town. So schools, tourist spots, religious places, travel centres, bus and railway stations, hotels, guest-houses, etc., are all important locations to take up first. Their sanitation again involves so many authorities.

Thus, it is important that everybody including the local authority undertakes a part of the sanitation program affecting its own area. Our budget reflects our will to do things. We all get the sanitation we deserve. Even in Europe and UK sanitation was very weak until early 1900s.

7.6.1 The Slum Sanitation Problem

A problem peculiar to an over-populated country like India is the problem of slums and their insanitary condition. Indian cities have at least 20–25% of their population residing in slums and shanties. Metropolitan cities like Mumbai have 50–60% of their population residing in slum areas where water and sanitation are both very difficult to provide. In slums, water supply is mainly through stand-posts and sanitation mainly through provision of community toilets ("Sulabh Sauchalayas"). Garbage collection and disposal from slums is neglected. Efforts are being made to provide more sanitary systems in slums, but sanitation is still a long way off. An important problem is that the residents do not hold any tenure on the land and are, therefore, at the mercy of so-called slum-lords!

An arrangement recommended for use in slums and shanty towns is to continue providing the Sulabh Sauchalayas for fecal matter disposal.

However, it is recommended to provide hut to hut water connections (instead of stand-posts) and supply only 10–15 L/person/day so that they can use the water a little more conveniently and at whatever time they need according to their work programs. The huts would then need to be served by a wastewater disposal system which can be provided cheaply in the form of "shallow sewers" as done in Brazil and other places. More details would be available on internet.

7.6.2 The Speed of Sanitation and People's Expectations

Even if one begins with young people, there is a certain speed at which sanitation improves in a country. This inherent speed can be increased by doing various things we have identified earlier. More than the speed, the point to note, however, is that the level of sanitation expected by the people improves with the flux of time. Already in one generation in India, the sanitation aspiration of the people has increased tremendously. With one more generation to go, the demand for better sanitation will increase so rapidly that we will be caught napping!

The speed with which sanitation demand increases with time is exponential, especially in urban and semi-urban areas. In India, its rapidest increase will occur in the small and medium-sized towns and in the existing cities and their peripheral areas. And it will keep improving further rapidly with time. Rural areas, however, will need greater time and effort. Whatever we can do to speed it up should be welcome.

> If the ideas are very expensive, they won't get implemented. If the ideas are economical but not scalable, they will not work. If the ideas are economical and scalable but not sustainable, it is of no use. They have to be all three!
>
> Dr Sam Pitroda

7.7 INDUSTRIAL WASTE TREATMENT AND "ZERO" DISCHARGE FROM INDUSTRIES

Industrial wastewater treatment is beyond the scope of this book. It involves a detailed study of site selection, the sources of waste, the extent of water used and whether "cleaner technologies" are adopted or not. Moreover, waste treatment is itself a specialized subject. Hence, we will not take it up here. However, since this book is likely to be read by all engineers

(Civil, Electrical, Mechanical, Chemical, Industrial, Electronics, etc.) a few concepts we would like to take up are discussed below.

At the outset, one must remember an attitudinal difference between an industrialist and an environmentalist. A man in industry is generally happy if he can account for 99% of the raw materials he uses. The balance of 1% may not be of any economic consequence to him. But, to the Environment Engineer, 1% lost in the effluent means a concentration of 10,000 ppm (or 10,000 mg/L) of that substance. The viewpoint of the two persons is very different. For the industrialist, 1% loss may mean good work while to the environmentalist it may mean a big problem and expense in removing it from the wastewater.

Secondly, municipal wastewater is, in fact, relatively cheaper to treat than most industrial wastes on a volume for volume basis. However, municipal wastewater is often much larger in volume (depending on the population) thus needing huge funds which municipalities do not have. India is a populous but poor country and municipal wastewater treatment has for various reasons fallen behind – thus leading to heavy pollution of water courses like rivers, lakes, estuaries, coastal waters, etc., due to people pollution.

Additionally, severe pollution may occur at industrial wastewater discharge points and industrial pollution has to be controlled as much as possible by treatment before discharge. Usually, treatment can be quite complex depending on the type of waste generated in the factory. Mechanized form of treatment is generally preferred for industries as they already may have skilled staff for operating their factory machinery.

Along with treatment, reuse is often promoted so that water conservation also takes place besides pollution control. India badly needs water resources for its various activities. States like Tamil Nadu are already forbidding industries from discharging industrial effluents to water courses so that they are forced to reuse. The authorities want as much reuse as possible in industries. Reduce, reuse and recycle are the modern mantra guiding all wastewater treatment.

Reuse can be done in two ways: reuse in industry itself and reuse for land irrigation (see Chapter 8). For reuse in industry, newer membrane and other technologies are now available for use. Several instances can be quoted. Reuse in land irrigation is already well understood and quite possible to use where there is little danger of pollutants entering the food chain. Thus "zero" or almost "zero" discharge is possible to conceive of in many cases.

7.7.1 Observation

Figure 7.3 shows schematically an industry where reuse is maximized in the industry itself and the balance is used for land irrigation. Discharge to the adjoining river is at a minimum.

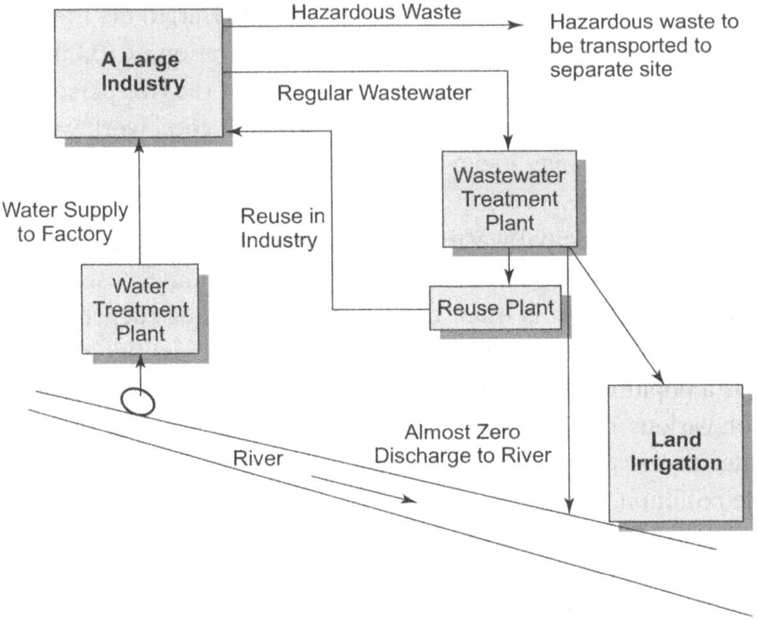

Fig. 7.3

Figure 7.3 Shows how almost "zero" discharge takes place from an industry to river. Reuse is done partly back in industry and partly for land irrigation. Discharge to river is thus minimized.

7.8 THERMAL POLLUTION

Thermal pollution in the environment is caused when an industry discharges an effluent that is warmer than the temperature of the water in the receiving body (river). The effect of this pollution on the ecosystem depends on the temperature rise which depends on the volume of effluent relative to the volume of receiving water with which it has mixed. As the river flows and the effluent mixes more and more with the river water, the temperature of the resultant mixture diminishes – and so does its adverse effect, if any. Thus, damage is always greatest at the effluent discharge point and decreases as the river flows. Therefore, effluent discharge standards only prescribe the maximum temperature difference allowable between the effluent and the river water, no matter what the volumes. Generally, tropical marine animals, molluscs, crustaceans are said to be unable to withstand a temperature increase of 2 or 3 degrees above their usual temperature and most are eliminated at temperatures above 37°C.

Figure 7.4 shows a high temperature effluent from an industry which needs cooling in a cooling tower prior to discharge to the river.

> Most standards prescribe that the temperature difference between the effluent and the river water (whatever their volumes) should not exceed 5°C. If the effluent temperature is higher than that of the river water by more than 5°C, a cooling tower or a cooling pond has to be provided to bring down the temperature within limits.

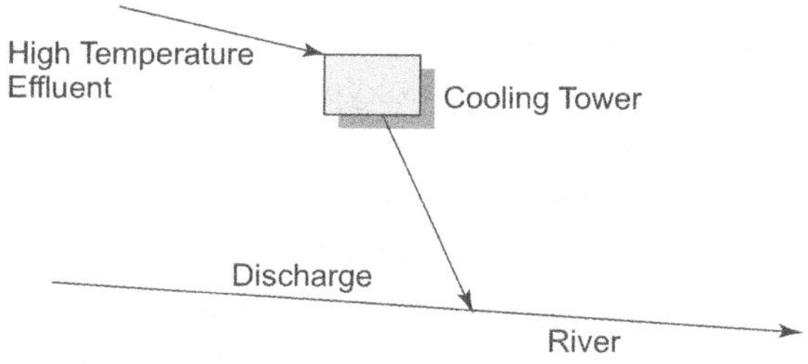

Fig. 7.4 *The provision of a cooling tower is often necessary to cool down an effluent before discharge to a river*

7.9 MARINE POLLUTION, COASTAL WATERS (CRZ), MANGROVES AND WETLANDS

7.9.1 Marine Disposal

Wastewater disposal in marine waters is not so simple as one may think! Firstly, there are several ecosystems involved. Secondly, the ecological effects that occur are complex. Things are further complicated by salinity, tides, waves, currents and such like phenomenon peculiar to a marine environment.

The typical elements of a wastewater disposal system to the sea are the following:

- ➢ Pumping and pre-treatment.
- ➢ Outfall to the sea with a diffuser at the end.

- The rising plume up to its natural level of entrapment, and
- Dispersion to the open sea as a result of currents.

The position of the disposal site is more or less fixed by the sewerage plan of the city or the location of the industry. Extensive site surveys need to be conducted before a system can be designed. As a rough indication, the extent of the survey area should cover about 1 km radius of the likely position of the diffuser and 500 meters on either side of the proposed outfall alignment. Much data are often available with the Navy and other institutions like the National Institute of Oceanography interested in the marine environment.

Water quality objectives to be met in case of sea disposal are essentially meant to protect people's health and well-being and promote the beneficial uses of coastal waters by minimizing any unfavourable impact on the micro-environment. Aesthetic problems, for example, may be caused by odors, the presence of floatables (oil, solid wastes) discoloration of water, etc., affecting recreational and navigational uses and adjacent land values. Other ecological effects would arise from oxygen depletion, excessive primary productivity (algal growth), reduction in water transparency, reduction in species diversity, and important health effects from bio-accumulation of pollutants in the fish food chain and/or affecting fish taste.

There are many coastal cities where municipal wastewater is properly disposed of by carrying it out into the sea through "outfalls" of adequate length and discharged into the sea either at the end of the outfall pipe or through a set of "diffusers" (e.g., Mumbai City).

Wastewater Treatment and Disposal Strategy

From the viewpoint of health and well being, care has to be taken on three counts when disposing wastewaters through outfalls: firstly, the outfall length and depth into the sea has to be adequate to meet the oxygen demand of the wastewater so that smell nuisance does not occur in the vicinity, secondly the people's health has to be protected from possible bio-accumulation of pollutants in fish and shellfish grounds, and thirdly, meeting coliform requirements at bathing beaches.

The Boomerang Effect

Degradable organics undergo bio-degradation in the sea in the same manner as they do in a river.

Fish foods both from estuaries as well as the open sea, are an important source of livelihood and nourishment for the relatively large coastal populations of India. Interestingly, fish act like a "boomerang", bringing back to the people on their dining table all the conservative and toxic pollutants the community vainly tries to throw out to the sea along with their polluted wastewaters. Such is the power of the food chains!

The Boomerang Effect

In the case of a river all the "problems" occur downstream from a city for someone else in the next city downstream to suffer whereas in the case of sea disposal it is the same community that suffers from its actions because of the boomerang effect. Any persistent pollutants (like mercury) thrown out to the sea come back to the community through the food chain (fish).

In assessing the environmental impact of a pollution control system for sea disposal, the proposed wastewater treatment facility and the outfall design have to be seen together in arriving at an optimum design. Both together achieve the end-result.

Some minimum degree of treatment is always given to all urban wastewater before discharge to the sea. Attention is always given to screening for the removal of any unsightly floating objects and to treatment for removal of oil and grease and the removal of grit-like solids which might settle in the outfalls at low flow and obstruct normal flow later.

All waste stabilization (BOD removal) occurs in the sea through natural biological processes. Some biological pre-treatment on the shore is considered only if it is cost-effective in shortening the outfall length as a result of this treatment. Advanced treatment to remove nutrients may be necessary in case of discharge to areas prone to nutrient build-up.

Persistent substances must be controlled at their source itself (in manufacture and use) as their control after discharge is practically impossible. From the point of view of bio-accumulation, persistent substances like mercury, cadmium, lead, chromium, arsenic, as also radio-isotopes, hydrocarbons, oils, PCBs, and pesticides like DDT, aldrin, dieldrin, malathion, from agricultural and other run-offs should not be present in discharges to the sea. Wastes, no doubt, undergo much dilution in a sea, but biological accumulation of persistent substances in the food chain may cancel out that benefit (again boomerang effect). Thus removal at source is the best, if not the only way.

7.10 COASTAL WATERS AND COASTAL REGULATION ZONE (CRZ)

India has a long coastline and some of its important cities are located on it. The Government is therefore keen to keep the coastal areas open to the common public and accessible to all. They have therefore framed some regulations controlling development in these areas referred to as the Coastal Regulation Zone (CRZ). Details are available from State Pollution Control Authorities.

Less than about 5% of the ocean area is involved in the continental shelf but it is the most important area from waste disposal viewpoint. It is also the area on which a large number of fisherfolk and allied personnel depend for their livelihood, and the general public for its recreation.

City waterfronts lend character to a city. They fulfill a variety of functions: commercial, transportational and social. Many a picturesque waterfront built with loving care in the old days has gradually turned into a sore sight with overcrowding, noise, pollution and such affronts to its original state. Improper land use is the principal problem in each case causing impacts of an aesthetic, social, physical or ecological nature.

In older developed areas, public roads were generally run along the waterfront. In newer areas, the road is pushed away from the sea thus depriving the public of a landscape amenity. Industrial air pollution problems can also affect waterfronts especially where industries are almost on the shoreline. Their presence at the shoreline can also cause effluent disposal, oil, smell, noise and traffic problems. For CRZ purposes the critical area is identified as follows. The mean sea level is the plane about which the tide level oscillates. Mean high tide level is the average height of all the high waters, and the mean low tide level is the average of all the low waters measured over several years. Land use in this area is controlled.

In India, the CRZ is further designated as I, II, etc., from the viewpoint of permissible land use in those areas and pollution potential. CRZ regulations generally prohibit construction activity upto 500 meters inland from high tide levels.

Bays and gulfs are a part of a sea or ocean forming a curved recess into the land. A bay is smaller than a gulf but is of the same general character. The Indian Ocean is an open ocean but the Bay of Bengal and the seas around the various Islands in the Ocean have current patterns and nutrients in varying concentrations. Estuaries exist where a river meets a sea and are important areas for habitation since history shows that worldwide they were generally the first to develop. Kolkata city is such a development. Moreover, estuaries support fish which move up a river for spawning and return to the sea at other times. They are thus susceptible to estuarine pollution. The famous Hilsa fish of Kolkata is a case in point.

In coastal areas of many countries including India, each river that empties into the sea washes down with it all the polluting wastes, nutrients and debris from the upstream land. Most of the heavy metals used countrywide ultimately are washed down to the country's coastline and deposited there. Similarly, all the drains (storm sewers) which carry storm-runoff from a city generally discharge into the near shore waters along the coastline. Such discharges also carry nutrients and other wastes washed into the storm drains.

In case of an over-populated city like Mumbai, overloaded sewers are "relieved" by allowing them to bye-pass sewage into the storm drains. Thus, even some raw sewage goes directly to the coastal waters via the storm drains instead of to the sewage outfalls and swimming in such waters is a positive hazard. BOD values of coastal sea waters normally range between 1.0 and 3.0 mg/L but in case of polluted shorelines (like Mumbai's) they may reach 30–40 mg/L on occasion. When traveling by air one often sees clearly in the case of Mumbai (and many other cities) a relatively bright greenish color of the algae-laden coastal waters just adjoining the land mass and the bluish color of the waters of the open sea.

Global warming is also raising the sea levels all around India (by about 4 mm/year) and all low lying coastal areas will likely show the health and other ill-effects of the same in due course. As sea level rises, saline water intrusion increases landwards and low-lying areas especially will get flooded. Well waters in coastal areas will turn brackish and agriculture will be affected in those areas.

7.10.1 Mangroves and Wetlands

Wetlands in general are the interface between land and water systems. Among fresh water wetlands one may include lakes, ponds and marshes whereas among coastal and saline water, wetlands one includes mangroves, swamps, marshes and coral reefs. Wetlands is a general term. Besides natural wetlands there are manmade ones such as the "Constructed Wetlands" we referred to earlier.

The Mumbai Metropolitan Region has been bestowed with large growths of mangrove vegetation that exists within the inter-tidal zones of estuaries and creeks in Manori, Thane, Dharamtar, Vasai and Virar regions covering nearly 5,000 ha of area within Greater Mumbai Region alone. India's total mangrove areas were reported to be about 700,000 ha in 1987.

Incidentally, mangroves make good sinks of carbon as they are reported to be able to absorb 50 tonne of carbon per hectare. Mangroves and wetlands are rich in biodiversity and consequently very productive. They support the livelihood of millions of people. Yet, they are also often the most endangered areas – with men destroying the wetlands to create dry lands to build upon. They have not always been managed well. From 700,000 ha in 1987 they had almost halved to 453,000 ha by 1995. Wetlands like many other resources have been up for grabs.

The mangroves are under constant threat of destruction and reclamation due to development pressures in the region. Builders are always looking for new land to build on. Government is itself sometimes guilty on the same count as they wish to create new land for roadways, railways, playgrounds, etc.

Concerns about declining fish stocks, increasing industrial and domestic waste discharges, and heavy siltation in Port areas have been voiced. Many threats exist to the mangroves in the region and government is expected to protect and promote the mangrove habitation in these circumstances. Fortunately, new mangroves can be cultivated along a coastline.

> Where public awareness is poor, mangroves tend to be neglected and get destroyed. Public vigilance is necessary to stop this decline.

Studies have been carried out from time to time on the West coast by agencies like MMRDA, The Bombay Natural History Society (BNHS) and the SP Godrej Marine Ecology Centre. Several studies have also been made by academic institutions, notably by the IIT, Powai, and a few other educational Institutes.

Several recommendations have been made from time to time on how to protect these mangroves as also on how to promote the growth of new mangroves. These recommendations have yet to become policy guidelines, and the help of the institutions named can be taken in this regard.

7.11 COMMON EFFLUENT TREATMENT PLANTS (CETPS)

Common Effluent Treatment Plants (CETPs) are essentially industrial waste treatment plants which serve a *group* of small and medium-scale industries, each contributing a relatively small quantity of effluent which is polluting or hazardous in nature and needs treatment which would normally be beyond the capability of each small individual unit to set up and operate skillfully. Large industries often prefer to have their own individual treatment plants and take full responsibility for operating them properly.

The objective of grouping small industries is to benefit from the resulting larger size of the operation so as to be able to afford appropriate technology through sharing of costs and skills. A large unit is often able to set up its own treatment facility as required and employ the right people to operate it.

This section has been included in this book as some of the engineers concerned may one day wish to set up their own small industry. They will then have to consider everything included in this book and select their site depending on whether they need to join in a CETP or not.

7.11.1 "CETPs" in India

CETPs seem to have found a special application in India. In 2010, there were over 100 CETPs in India (30% in TN State alone) to serve groups of small and medium-scale industries. Most of them served clusters of less than 100 units per cluster, but some had over 600 units within a cluster. Flows varied from 200 to 55,000 cu m per day.

> Some CETPs serve *homogenous* units (e.g., all tanneries). Some have *heterogenous* units (various chemicals) and each factory runs production campaigns at different times of the year. Thus, the effluent is constantly changing, and a very diverse lot of problems is created. Correspondingly, a lot of experience exists countrywide in dealing with the problems.

7.11.2 How Relevant are "CETPs" ?

Of course, they are very relevant, locally. They have localized importance. CETPs serve only small and medium-scale industrial units grouped together for ease of treatment in a common treatment plant. Sometimes, the manufacturing units are themselves voluntarily relocated there in order to benefit from grouping. Another form of benefit occurs when one industry uses the wastewater of another industry nearby.

On a countrywide basis, the importance of CETPs is relatively small because industrial wastewater is only 10–15% of total wastewater produced in India. This is not to belittle CETPs, but just to put things in perspective. The remaining 85% is people's sewage. We are a very populous country and have a very large backlog of plants to be built just for treating people's sewage!

7.11.3 Some Thoughts on "CETPs"

It is best to promote the use of "clean technologies" at the planning stage for all types of industries—large, small and medium-scale—whether having an individual or a common treatment facility. To ensure sustainability Clean Technologies have to be followed by some Social Engineering (see Section 11.6).

As can be seen from the following Table 7.2, CETPs are often cheaper than individual treatment plants. Why? Because, generally CETPs have the advantage of making the wastewater more easily treatable and the treatment more affordable because of common operating staff looking after the plant. Of course, as stated earlier, for any specific waste it pays to reduce the volume of the effluent.

Table 7.2 *Costs of CETPs and a few other individual treatment costs*

Source	(₹ per cu. m. per day)
Cotton Textiles (excluding Denims)	5,000–8,000
Pharmaceuticals	20,000–30,000
Bulk drugs, antibiotics	50,000–90,000
Petrochemicals	80,000–120,000
CETPs	5,000–40,000

*Land cost is not included in above estimates

In the following text, we refer only to those items of a general nature in which all persons including NGOs and industry-owners would be interested.

7.11.4 Collection of Wastewater

A question emerges: Should collection be done by pipelines from each of the units contributing to a CETP or the collection done by tankers? It is often true that when a common pipeline is used, there is some scope for cheating by not declaring the true nature or quantity of the waste. Where tankers are used, better control can be exercised as samples can be withdrawn from each vehicle as it arrives. Many arguments seem to be valid for both sides.

Merely on economic basis, an early study showed the breakeven point at 2 km. Up to 2 km, both types were equally costly. For longer distances, pipelines worked out cheaper than tankers. Today's costs are different in most cases and a study of relative costs may need to be redone to arrive at an answer in a specific case.

7.11.5 Pre-treatment at Factory Level

Another question that emerges is the extent of pre-treatment of the wastewater that is necessary to be done at the factory site before the wastewater enters the common treatment facility. Pre-treatment is essential for certain items only (like oil and grease) at the factory level. Rest would perhaps be better done in admixture with other wastes at the common plant. Guidelines for pre-treatment are often reported to be too strict for small-scale industries which usually have weak in-plant controls. This is a problem area in some cases.

Correct flow estimation from the units is another problem area. Reliable figures are often difficult to get.

Operation

Experience shows that operation of CETPs is only as good as their operators and Government's enforcement policies. Hence, use qualified, certified, trained and experienced operators. Pay a little more if necessary.

On the monitoring and enforcement side, one is advised to tone up controls because if you let people cheat, they will cheat.

> With common as well as individual wastewater treatment plants, one must remember that monitoring and enforcement by the local pollution control authorities are essential to keep people in check.
>
> If you let people cheat, they will cheat!

REFERENCES

Arceivala, S J, and Asolekar, S, (2007), *Wastewater Treatment for Pollution Control and Reuse,* Tata McGraw Hill, India.

Recommended Additional Reading

The reader is requested to refer to the internet for more references and select those which seem suitable for additional reading.

ILLUSTRATIVE EXAMPLE

1. Prepare a preliminary plan for treating the wastewater from a fast developing town called "Nature-gaon". Wastewater treatment for the new areas has to be done using "natural" treatment systems (no electric power, no CO_2 emission) or by the simplest mechanized systems like aerated lagoons followed by groundwater recharge to conserve water and nutrients. You are required to determine their land and power requirements.

   The details are as follows:

   The old town with 200,000 population to the north-west of the lake is served by an old mechanized treatment plant (STP) which does not function well and discharges the half-treated effluent to a river as shown. It is proposed to replace the old plant by an Aerated Lagoon (AL) with floating aerators which is the simplest type of mechanized system operating on electric power. Its effluent is later proposed to be given additional treatment (at industry cost) to make it fit for reuse in the new industrial area on the west.

   A new *intercepting sewer* is also proposed to be laid along the upper part of the Ring Road to carry to the above treatment plant whatever wastewater reaches the sewer, while the lake would be protected against pollution.

Area	Population	Proposed Treatment	Remarks
Old town, North-west of Lake	200,000	Aerated Lagoon with floating aerators	To replace old, poorly working mechanized treatment plant and be capable of further treatment for reuse in industrial area on the West, if desired

North-east of Old town	5,000	Constructed wetlands	Effluent expected to replenish water lost from lake by evaporation. This will keep it topped up in summer and discourage misuse of lake fringes by encroachment.
Mixed commercial and residential area South of ring road	50,000	Oxid Pond + Irrigation	Water will be conserved and crops grown.
Private plots on South of ring road	700	Septic tank + soak pits	Water and nutrients conserved.

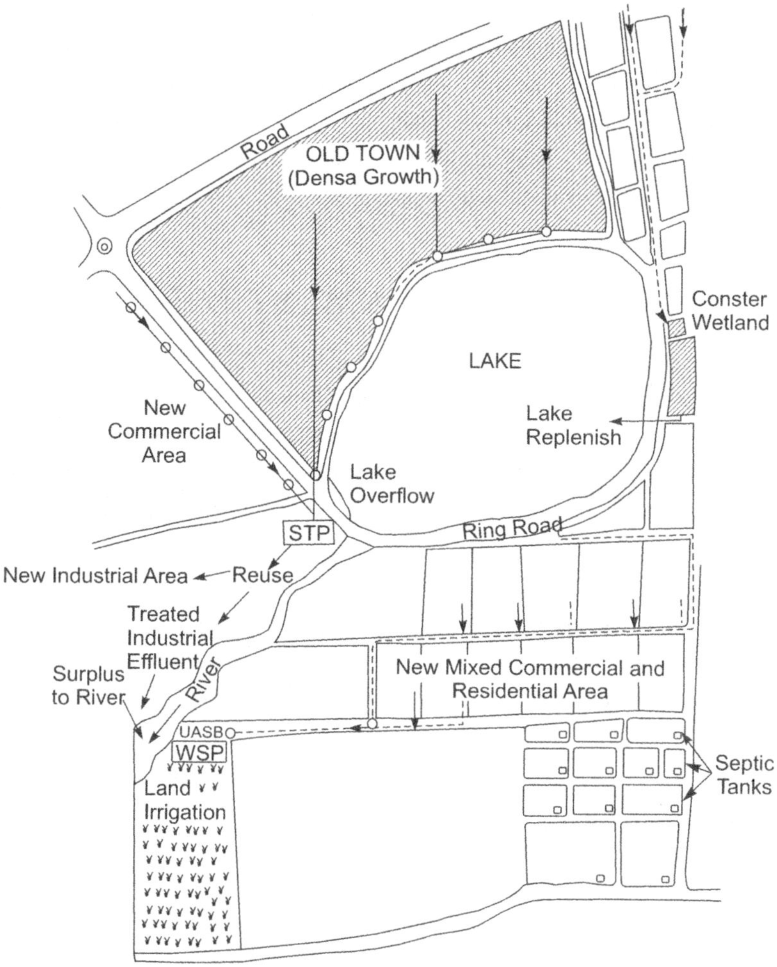

Fig. 7.5 *A decentralized wastewater disposal system planned for a newly developing area adjoining an old town*

Thus, a mix of 3 or 4 different natural treatment systems along with simple aerated lagoon for the old town (with scope for further treatment for reuse in new industries) are proposed to be used instead of just one centralized sewerage system discharging poorly treated effluent to the river, consuming expensive electricity and emitting CO_2 to the atmosphere.

Moreover, the overall cost of treatment would reduce as electric power would not be used by natural treatment methods. The lake would be protected and reusable water would become available for industry.

NOTE: The whole process would be like what the cinema people do: converting an old style single-screen Cinema house into a thriving new 4-screen multiplex better serving different needs, and hopefully making a profit. This could be called "affordable and sensible sanitation for all".

PROJECT/ASSIGNMENT TOPICS

1. It is desired to recycle wastewater from IIT campus. The objective is to treat about 75% of the total flow to quality needed for toilet flushing and gardening (assume that both uses need about the same quality). The remaining flow should be converted to potable water standards. How would you go about doing it? Describe your proposed treatment strategy, and record your assumed values of quality parameters for inlet sewage and the two treated streams?
2. Briefly discuss each of the following options for implementation, marking those which you think are more feasible to adopt, and those which might possibly have undesirable side-effects.

Soft Options

- Conserve use of water wherever possible as it also reduces wastewater to be handled.
- Hold seminars and discussions on wastewater treatment methods.
- Adopt natural or mechanized systems depending on land availability.

Hard Options

- Plan for standardization of wastewater treatment for rapid increase in coverage.

- Promote facilities for manufacturing standard sizes of floating aerators, sewer pipes.
- Promote consultancy services for industrial wastewater treatment.
- Implement a program to train the trainers for providing sanitation.
- Reuse where possible to strive for zero discharge.
- Secure Governmental / International Funds for system coverage / expansion.
- Attend to thermal pollution, marine pollution, CRZ, mangroves and wetlands.
- Plan measures to protect habitation in coastal areas against sea level increases.
- Set up CETPs wherever necessary to promote small-scale industries.
- In funding, use BOOT and such systems.

8 CHAPTER

Wastewater Reuse for Irrigation, Industry and Community Uses

8.1 Reuse of Wastewater in India

8.2 Reuse in Irrigation, Soil-Crop-Water Balance, Health Hazards

8.3 Reuse in Industry

8.4 Reuse for Augmenting Public Water Supplies

> *As everyone knows, necessity is the mother of invention. Likewise, scarcity is the mother of water conservation and reuse. Encourage water conservation and reuse at all times. To use Swami Vivekananda's words: "It is the divine whisper of tomorrow".*

8.1 REUSE OF WASTEWATER IN INDIA

Reuse of wastewater has been done in India for various purposes for over a 100 years now and surely will be done much more in the future. Reuse has also picked up speed and promises to become an important requirement of the future to conserve water. Four distinct steps can be traced in its development in India:

1. Reuse in agriculture and horticulture, from 1895 onwards.
2. Reuse in industry to meet chronic water shortages, from the early 1960s.
3. Reuse in industry to meet product recovery, pollution control and other such objectives, from the 1970s and 1980s.
4. Reuse for augmentation of public water supplies, from 1990s onwards.

The cost and complexity of treatment has increased with each grouping as we see below.

Salts Picked up by Water in Usage

The original water is usually of drinking water quality. However, during use it picks up various substances and becomes wastewater which has to be re-treated to give reusable water. The mineral pick-up from domestic usage of water is given in Table 8.1 (from US experience).

Table 8.1 *Mineral pick-up from domestic usage in USA (Metcalf & Eddy, Inc. 1990)*

Item	mg/L
Chlorides	20–50
Sulphates	15–30
Nitrates	20–40
Phosphates	20–40
Sodium	40–70
Calcium carbonate	15–40
Boron	0.1–0.4
Total dissolved Solids	100–300
Total Alkalinity as $CaCO_3$	100–150

Treatment Depends on End-Use

Treatment for reuse depends on the end-use proposed to be made. It may include all three forms: primary, secondary and tertiary treatment or to put it simply, one may have to provide treatment in three familiar steps – first, water treatment, second wastewater treatment, and third industrial water treatment – to meet the final requirement.

8.2 REUSE IN IRRIGATION, SOIL-CROP-WATER BALANCE, HEALTH HAZARDS

It has been appreciated worldwide that wastewater irrigation is not a primitive or inferior method of disposal, but rather a logical one to use where the climate and local soil conditions are suitable. In fact, it gives full primary, secondary and tertiary treatment – all rolled in one! Germany first started irrigation with sewage in 1880 in Europe.

In India, sewage disposal by irrigation has been the favourite method of disposal since almost the same time. India with its warm climate has been practicing land irrigation since 1885 and presently reportedly has over 150 farms covering over 12000 ha of land. Agricultural reuse of municipal wastewaters has been practiced in India since 1895, but for all the wrong reasons:

- ➢ Municipalities want to get rid of their untreated or "half-treated" effluent.
- ➢ Farmers are happy to use it for its water and nutrient value.
- ➢ Both neglect treatment and health aspects.

Actually, three things occur when any wastewater is reused in agriculture: water and nutrients are reused and soil conditioning occurs. Charges for reuse in India are generally based on its water value alone, and amount collected is credited to the general Government coffers. The staff, who operate the treatment plant is not benefitted by operating it well. This is discussed further in Section 11.7 while discussing sustainability.

In 1980, USA had 1000 municipal plants and 1300 industrial plants using irrigation. In USA, irrigation is compulsory to consider first as a method of treatment where tertiary treatment of wastewater is needed to meet discharge standards. In USA, however, irrigation is not permitted for food production unlike in India where irrigation is mainly done for food production. Therein lies a big difference.

Design of Wastewater Irrigation Systems

If a large irrigation facility is proposed to be set up, it is recommended that an agricultural expert be associated with the civil and sanitary team handling the waste flow and treatment aspects.

Wastewater irrigation is best for treating domestic and similar wastewaters. It is also considered most appropriate to use with agro-industries and food and beverage industries wastes. Thus, a large number of wastewaters such as the following are most suitable for irrigation:

- ➢ Domestic sewage.
- ➢ Industrial wastes from food and beverage industry.
- ➢ (Canneries, milk dairies, sugar factories, beverage factories breweries, distilleries, etc.).
- ➢ Effluents from fertilizer factories, Pulp and Paper industry wastes, etc.

Use of domestic wastewater for irrigation, as stated earlier, is often done for the wrong reasons. The municipality wishes to get rid of its wastewater without any complaints from the pollution control authorities knowing that it has not been treated up to the mark, whereas farmers are willing to take over this wastewater for land irrigation because of its water value knowing that partial treatment is good for them as organic matter left untreated in the wastewater is good for the soil and the crop. So, poor compliance is good for both!

Wastewater irrigation is not recommended for wastewaters containing hazardous substances such as:

- ➢ Oils, tars, petroleum products, solvents, etc., which tend to migrate in the soils and spread far and wide through the ground waters.
- ➢ Heavy metals and various toxic substances from industries which accumulate in the food chain and affect health.

Agricultural reuse of industrial wastewaters is sometimes fraught with danger as it may contain toxic and/or accumulative pollutants which can be spread through the food chain and affect the health of consumers. Quality of industrial wastewaters used for irrigation by an industry on their own land often escapes external control and sustainability of the land may often be gradually affected for the worse. Industrial wastewaters suitable and unsuitable for land irrigation are described further below. The financial policies of funding agencies either promote or discourage wastewater irrigation according to national policy in this respect.

Land Requirement for Irrigation

Firstly, the volume of wastewater available for irrigation is estimated and its average, minimum and maximum values determined over the seasons. In India, land required for irrigation is estimated roughly @ ~100–200 cu/m/ha/day approximately. When flow is at the maximum rate, either additional land should be available for use at that time or the excess flow has to be handled separately before discharge to a near-by water course.

Crops to Grow

Ideally, wastewater quality should be such that it can be applied to the soil in the required quantity without causing any problems for the soil and the crops. Hence, quantity and quality should both match. It is often best to be guided by local practices in determining what crops to grow. Usually, choice of crop lies between the following:

- Forage crops (fodder grasses, alfalfa, etc.)
- Field crops (Wheat, barley, etc.)
- Oil-bearing plants (Citronella, Mentha)
- New bio-fuels (Ethanol, Jetropha, etc.)
- Fruits, flowers, ornamental plants

In this day and age, oil-bearing plants and bio-fuels seem attractive to grow, but sometimes it is more important to use the land for growing food crops which the people need.

The Indian Method of Pretreatment

Municipal wastewater generally has a BOD ranging between 250 and 400 mg/L. The method preferred in India is to first treat municipal wastewater to make it fit for land irrigation so as to be able to cultivate food crops. Thus, treatment is usually done to reduce BOD from its original value to about 70 mg/L or thereabouts which is considered enough for use in irrigation. The effluent after irrigation is generally considered fit for discharge to a local water course (stream or river).

This degree of treatment (~400 → ~70) can be obtained by use of aerated lagoons. The wastewater may be sent to an aerated lagoon of required size so that the effluent can be used directly for irrigation. However, aeration requires electric power at site. Where power is not available, it can be generated at site using a diesel generator or solar panels. Interestingly, solar energy is now reputed to be often competitive in cost with diesel power when all additional costs are taken into account (see Fig 8.1).

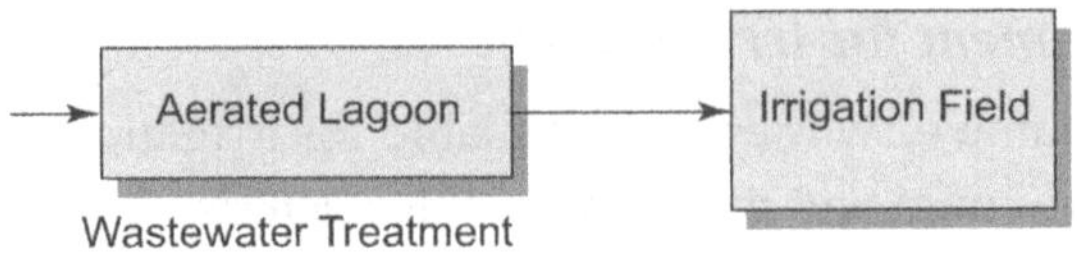

Fig. 8.1 *Municipal wastewater treatment in aerated lagoon prior to irrigation*

> After all, irrigation is the most preferred way for disposing municipal wastewater in the warm climate of India. It also helps produce more food for the people.

Where there is power shortage or where the wastewater BOD is high (upwards of 500 mg/L) the treatment method recommended is one that requires no electric power. One can then use the UASB and, in fact, produce biogas which can be converted to electric power, if desired. The anaerobic effluent from the UASB may be further treated in a short-detention oxidation pond rich in algae. Both treatment methods can produce an effluent fit for irrigation.

Much resource recovery can also occur in the form of reusable water, biogas, and other edible food products such as crops, fish, etc. Generally in India, no balancing reservoirs are provided before irrigation (as in US or Europe where winters are too cold for irrigation) (see Fig. 8.2).

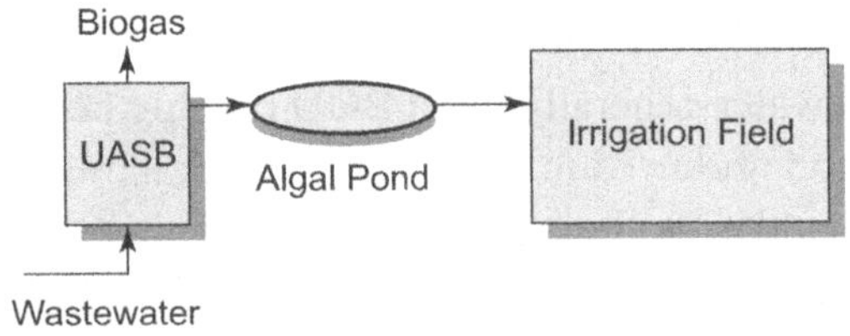

Fig. 8.2 *Wastewater pre-treatment in UASB for high-BOD wastes prior to irrigation*

Soil-Water-Crop Balance and Food-chains

Repeated irrigations are needed during the crops' growing season. As is well known, the water applied in each irrigation undergoes losses in three ways:

- Gravity loss
- Hygroscopic loss, and
- Evapotranspiration loss

A part of the irrigation water applied just percolates underground owing to gravity action within the first 1 or 2 days. Another part of the water is held hygroscopically by the soil and does not move. A third part is available to the plant for growing by photosynthesis during which water is lost by evapotranspiration. Thus repeated irrigations are necessary as a plant grows.

Each time water is lost by evapotranspiration, only the water evaporates while the salts contained in the water are left behind. Thus, the quality of the soil changes slowly with time and salt build-up occurs in the soil.

When it rains, the salt concentration built-up in the soil washes down and decreases. The extent of wash-out depends on the extent of rainfall and the type of soil. A sandy soil can hold no water while a clayey soil holds a lot of water. Most soils are somewhere in between the two extremes. Soil and water have to be analysed and an agronomist is needed along with engineers when planning large systems.

From the viewpoint of wastewater disposal, it would seem to be best to grow various types of grasses and fodder plants as they have a relatively high evapotranspiration rate and, therefore, need more frequent irrigations. At the same time, grasses are more tolerant to the salt build-up which must occur in such cases. Thus, wastewater and grasses seem to be made for each other! Suffice it to know that the soil, water and crop have a relationship between them and the selection of one affects the other in course of time. This subject is discussed more fully elsewhere (Arceivala & Asolekar, 2007).

Furthermore, heavy metals and toxic materials tend to pass from the water to the soil and thence to the crop, thus accumulating in the food-chain and affecting the health of the animals that graze on it. From the animals the material can pass via the milk and the flesh to human beings. Both vegetarians and non-vegetarians can suffer.

Lysimeter Tests for Soil Treatment

Soil is an excellent medium for waste treatment and is particularly helpful when reuse is intended. The earth contains millions of small particles of soil with a large volume of interstices between them. If the particles are large (e.g., sand) little water can be held in the interstices as they drain away by gravity leaving behind only a sandy desert. The smaller the particles (e.g., clay) the more the moisture that can be held in the interstices but the greater the hygroscopic forces which do not let them part with the water.

When wastewater is applied on land, it percolates down coating the millions of tiny soil particles with organic matter and filling the interstices too. The soil is also rich in microbial activity throughout comprising the three sets of decomposers, namely, aerobic, facultative and anaerobic micro-organisms. Their activity continues using first the free available oxygen in the interstices after which breakdown of nitrates and sulphates begins as explained in Section 7.1 and 7.2. When sulphates are broken down to sulphides, foul smells are caused and the soil is referred to as "sick" soil.

> Each soil has a certain capacity for microbial "treatment". If overloaded, it becomes anaerobic and foul smells are caused by the sulphides formed. This is called "soil sickness". Natural earth is therefore an excellent treatment device, but only up to a point.

When soil is loaded in special standard cylinders called *lysimeters*, they can be used to determine how much organic load a soil can tolerate. This can help determine the extent of pre-treatment necessary to be given to a wastewater so that the treated wastewater when applied to a soil would show aerobic conditions throughout and allow the plant to grow. One may study actual growth of plants and corresponding soil conditions under controlled conditions in "lysimeters" operated over a few months.

Especially in case of certain industrial wastes, typical lysimeters can be used for actual experimentation. A lysimeter would be like a short piece of pipe filled with soil in which a plant can grow. Several lysimeter tests using the local soil may need to be conducted in parallel to determine the plant growth and soil conditions at different wastewater application rates (an application rate of 1 cm/day in the test lysimeter = 100 cu m/ha/d on actual farm). Tests are carried out over several months (FAO/UNESCO1973) (see Fig. 8.3).

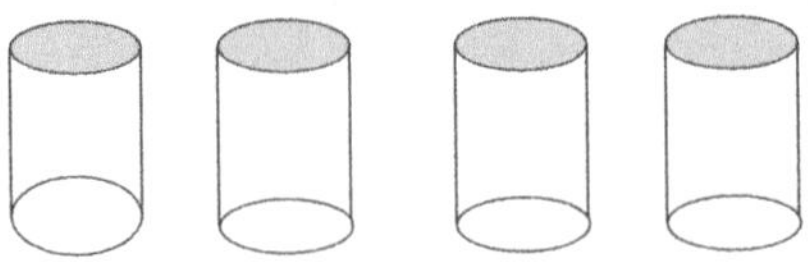

Fig. 8.3 *A lysimeter test with four units set-up in parallel to find acceptable loading rate*

Example: For determining the extent of pre-treatment necessary for a pulp and paper mill wastewater proposed to be disposed by irrigation in India, a number of lysimeter pots were set up.

The optimum loading was determined from lysimeter tests to be 225 kg BOD_5/ha/day since at that loading the growth was found to be relatively better and no anaerobicity was found in the bottom of the pots. This information helped to determine the extent of pre-treatment necessary to be given to the waste so as to reach the indicated value. The treatment plant was designed accordingly.

Health of Farm Workers

Everyone must be aware of one more problem in using irrigation for wastewater disposal. The health of farm workers has to be protected from hookworm infection during operations as they are exposed to it. Regular medical check-ups are necessary besides wearing of boots and gloves to prevent hookworm infection from the soil to the worker.

An "Environmental Impact Study" is first carried out to set the baseline standards and ensure:

- Good engineering design of components
- Safe handling of wastewater (i.e., without human contact and with protective clothing), and
- Provision of proper flow controlling and flow measuring stations, etc.

These features set apart a well-designed and well maintained wastewater irrigation farm from the rest of the crude efforts made by people to benefit from irrigation while neglecting its health effects. In any wastewater irrigation system one must remember that one is dealing with a kind of "mixed blessing" situation. The wastewater is a mixed blessing indeed as there is its useful water-value and its important nutrient-value both of which come with its dangerous chemical and microbial aspects which can only be neglected at ones peril.

> The objective in sewage farming is to run a *safe and sustainable* wastewater disposal project which is firstly, safe at all times for the health of the workers in the treatment plant and in the irrigation field, and sustainable from the viewpoint of soil condition. Otherwise, the crop output may diminish in course of time and nutrient value of the crops may be depleted.
>
> The impact on the quality of the surface and underground waters around the site is also important.

8.3 REUSE IN INDUSTRY

Reuse in industry has been done in India for various reasons as shown below.

Industrial Reuse to Meet Chronic Water Shortages

Plain water conservation Reuse of wastewater in industries to meet chronic water shortages in India first began in Mumbai, in early 1960s when shortfalls in city water supply became pronounced. Reuse began first with plain conservation and progressively developed into more and more complicated methods, as described below:

This came first in the textile industry in Mumbai where it was shown that 15–20% water could be reused directly without any treatment. One had to just find the manufacturing steps from which the wastewater could be held in a tank and recycled in other processes where quality was not important. Higher percentage of reuse was possible if the wastewaters were pre-treated and then reused.

Thereafter, many other industries were surveyed for similar water conservation possibilities. Maximum reuse potential existed where cooling waters were involved.

Tertiary treatment and reuse of sewage at individual building level Several commercial buildings which suffered water shortages also showed that reuse could be done using the building's own toilet waters. The toilet waters were given tertiary treatment in the basement of these buildings and the treated waters were pumped back for reuse as make-up waters for cooling towers of centrally air-conditioned buildings. Provision of centralized A/C thus became possible for the whole building in case of prestigious buildings in spite of water shortage, like in the Air India Building at Nariman Point, Mumbai, where it was first done in early 1960s. Around 20–30 plants of this type then followed in India (including 5-star hotels like Taj, Oberoi, etc.)

Operating costs worked out to ₹ 5–8 per 1,000 litres (1985 costs). This made reuse water cheaper than fresh municipal water, at that time, and more assured. Some treatment plants were renovated to replace worn out machinery retaining the old processes which worked well.

For larger consumers (factories), sewage was taken from municipal sewers and tertiary treatment given by the industry in question, on an individual basis. This was again done first in Mumbai in 1968 with the

Union Carbide factory at Chembur (which later became Oswald Agro). (see Fig. 8.4).

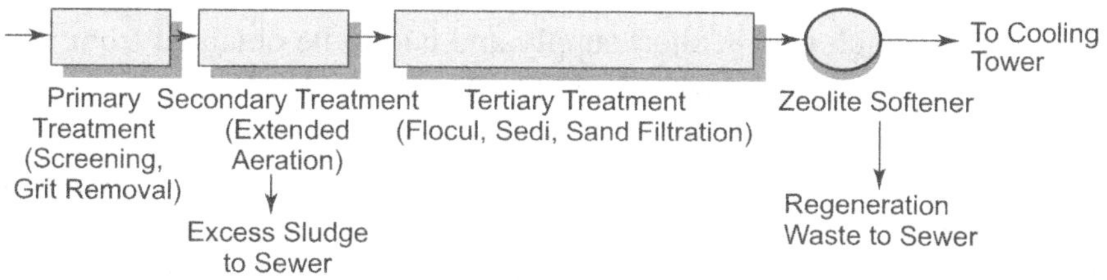

Fig. 8.4 *Typical flowsheet for treating wastewater for reuse as cooling water*

Years later, two more tertiary treatment plants came up in Chennai for Madras Refineries and Madras Fertilizers, using more concentrated Chennai sewage and later one more plant came up in Mumbai for Rashtriya Chemicals and Fertilizers (RCF) at Chembur. They all gave reusable water at costs ranging from ₹ 12 upto as much as ₹ 35 per 1000 L (1995 costs) depending on the nature of the original wastewater and the degree of treatment required.

In the US, reuse is done on a citywide basis (see later). This requires laying two separate distribution systems, one carrying potable water and the other non-potable water. Such duplication of the supply network is not possible in India where the streets are relatively narrow and traffic heavy. In India, reuse is done on an individual basis with individual funding.

Industrial reuse to meet other objectives It was shown that:

1. **Reuse could be done at little extra cost** over that required for regular pollution abatement, but it gave large savings in cost of fresh water saved through reuse. For example, the effluent from a milk dairy could be treated for use as an irrigant to grow fodder for the milk cattle and to help in initial bottle washing operations thus saving some fresh water.
2. **Reuse of water helped resource recovery** for the industry (like recovering paper fibres from paper mill wastewater) and saved the cost of abstracting fresh water from the ground or river and paying the corresponding "cess" charges payable to Government for discharge of wastewater.
3. **Reuse reduced dependence on river flows**, and
4. **Reuse helped industries to gain tax advantages** from certain locations in arid or designated zones reserved for *zero* discharge of effluents.

An extremely interesting example of reuse in India is the recent case of Tirupur, Tamil Nadu which has several industries engaged in manufacturing and exporting tee-shirts, etc., to the Western countries. They needed clean water which was in short supply and had to be obtained from long distances by use of expensive tankers. Moreover, the dyeing operations generated coloured and salty wastewater which ruined the surroundings. Their solution was to reuse the wastewater after a high degree of treatment using membrane technology. This solved both their problems. Capital funding problem was solved by having the plants installed at the cost of the vendors on "Build-Operate-Transfer (BOT) basis".

8.4 REUSE FOR AUGMENTING PUBLIC WATER SUPPLIES

This kind of wastewater reuse is only now beginning in India. Its potential for "value addition" is the highest. Whole new water resources can be generated in this way for the benefit of a community by using treated *wastewater* to augment the regular water supplies! Fantastic possibilities exist! Like water harvesting, we can call it "Wastewater Harvesting". The reclaimed water could be reused for a variety of purposes, such as the following:

(1) Ground Water Recharge

Where groundwater levels are falling fast (e.g., Delhi) recharge of the aquifer can be done with treated wastewater, the reclaimed water, so as to bring up the aquifer levels and reduce pumping costs for the whole community.

(2) To Control Saline Water Intrusion

In fast growing coastal areas on the West coast of India, sea water tends to creep in and replace fresh water which is abstracted out. This makes coastal water more brackish with time (Goa, etc.). Here, wastewater is treated and the reclaimed water is pumped into the ground to control sea water intrusion.

This method will be required increasingly in the future, perhaps. Global Warming will increase sea water levels which in turn will increase sea water intrusion which will need to be controlled to save the coastal habitations.

(3) Reclaimed Water can be Used for City Supply

Reclaimed water can be used for augmenting the city supply by providing two distinctly separate water supply networks in the city as is done in some US cities

(a) Normal potable quality supplies

(b) Non-potable quality supplies

USA has been using dual supplies since 25 years: [California (57 MLD), Florida (85 MLD)]. US experience: All trace chemical contaminants may not be removed from reclaimed water during treatment. Yet, no problem if *occasional*, inadvertent ingestion by public takes place.

However, this kind of system may have its own problems in a country like India. Gardeners, servants, etc., may drink non-potable supplies regularly, out of ignorance. Also, laying a new network of pipes in an Indian city would cause severe traffic problems with dug up roads.

New Delhi is said to be considering dual supplies in selected areas where aquifer levels are dropping fast. Generally, India prefers to reuse reclaimed water on an individual basis than on community basis, as described earlier.

(4) Recreational uses of Treated Wastewaters

Wastewater can be treated and reused for recreational/touristic purposes such as the following:

- Creating a new lake for recreational purposes (body-contact sports?)
- Boating and (non-contact) water sports
- Bird sanctuaries, Livestock watering
- Wetlands development

Augmenting Public Drinking Water Supplies in India

> In general, Indian lakes are different from Western lakes. We lose 30–40% of lake volume per year due to evaporation during the dry season of 8 months. Hence, every lake in India actually needs some replenishment.

Spreading chemicals to control evaporation has not been very successful in India.

As explained in Section 6.4, Hussain Sagar lake in Hyderabad is the first lake in India to be supplemented by treated wastewater (20 MLD) from a residential area upstream of the lake to supplement, by gravity, the natural catchment inflow into the lake so as to make up for high evaporation losses in its warm climate. This has been possible to do since 1998. The lake is not used for potable purposes. Supplementation keeps it "topped up" and ensures overflow each year. It controls anaerobicity, smell nuisance, controls build-up of nutrient in lake and protects other beneficial uses. As stated earlier, lake is used for *non-potable* uses only.

> Supplementation with treated sewage can be done even where a lake gives *potable* supplies, provided the tertiary treated wastewater is given further treatment by a natural process such as groundwater recharge. This is what is conceived of at present in a case in Jaipur, Rajasthan, and soon it will be proved by practical experience.

There are two ways in which reuse is being done in other countries:

1. Direct reuse for potable purposes can be done as in Windhoek, Namibia, South Africa, where it is being done for the last 25 years. Well-treated wastewater is blended with fresh river water on 50:50 basis and supplied for direct use in town). As stated, this form of direct reuse is being practiced for the last 25 years!
2. Indirect reuse for potable purposes is already being done in several countries (USA, Europe, Israel). In these countries, wastewater is first well treated. Thereafter, groundwater recharge is done. In flowing underground, its "identity" is lost. It is then re-abstracted and supplied either directly (USA, Israel) or after a second round of treatment is given (Holland).

Singapore is now beginning to reuse. They have no land for lakes or for tapping groundwater. They cannot depend forever on Malaysia for a part of their fresh water supplies. Hence reuse of wastewater after treatment (using membrane technology) or desalination of sea water are the only two approaches they can have recourse to.

> Besides desalination, we have the technology for renovating wastewater into potable water. But, the question is its acceptability. Just how far are we prepared to reuse?
>
> For a country like India with a large population, water resources will never be enough and treated wastewater will someday have to be used even for replenishing "potable" water supply lakes. This is a price we have to pay for our large population. Let us do it well.

US, Europe and Israel are already doing it in certain areas. Singapore has started reuse. In most cases, the wastewater is first treated well in a treatment plant and then made to flow underground for "natural" treatment. Its quality is further improved, its original identity is lost, and its reuse becomes acceptable. The ground water may have to be treated once again before supply. We have to use our reuse expertise to overcome the heavy evaporation losses which occur from our Indian lakes. In this way, we will have turned our tropical climate to our advantage, and will still have a few lakes left to call our own!

REFERENCES

Book entitled: *"Water Renovation and Reuse"* (Ed. Hillel Shuval) Academic Press Inc., 1977. Chapter entitled "Water Reuse in India" by S J Arceivala.

Arceivala, S J, (1981), *Wastewater Treatment and Disposal*, Marcel Dekkar, USA.

Arceivala, S J, and Asolekar S, (2007), *Wastewater Treatment for Pollution Control and Reuse*, Tata McGraw Hill, New Delhi.

Arceivala, S J, Kapadia, J R, Wadekar, V R, (1969), "Economy and Reuse of water in Textile Mills", Proceedings All-India Textiles Conference.

H B Singh, (2003), Personal Communication.

AIC Watson (1995) Report on Hussein Sagar Lake submitted to Hyderabad Metropolitan water and Sewerage Board, World Bank Project 1993–1995.

PROJECT/ASSIGNMENT TOPICS

1. List the various ways in which reuse can be done. Which way appears to you to be the cheapest, and why?
2. You have been chosen as a consultant for setting-up of water recycling facility for the Municipal Council of Timbuktu. Explain how you would handle the water and wastewater situation given the following information:

   As you are aware, the City of Timbuktu is situated on the shore of a polluted lake called "Timbuk Sagar". As it stands today, the Timbuk Sagar is polluted due to deposition of Timbuktu's wastewater (approx. 10 MLD flow) over the period of past two decades. Now, that the life of Nano Auto Factory of Tata's has been made miserable in the State of West Bengal, Tata's would like to relocate their factory. Clearly, nobody can compete with the City of Timbuktu because the City is blessed with two lakes i.e., drinking water-quality lake (Nirmal Sagar) and above mentioned Timbuk Sagar.

Much of the water that comes to the Timbuk Sagar is the partially treated (mostly primary treated sewage) 10 MLD flow. The present population of Timbuktu is aprox. 80,000 and draws aprox. 10–11 MLD water from the Nirmal Sagar; which is further treated in a drinking water treatment plant (called as water works) and distributed in the City to fulfill the daily average requirement of 135 L potable water per person per day. Presenly, the City enjoys patronage from three major academic institutions and some trading activity of agricultural produce. With Nano Auto Factory coming to Timbuktu, there is going to be the addition of around 30,000 new population from the Auto Factory and probably another 10,000 population from service industry such as restaurants, lodges, and Apana Bazaar, etc. The hydrological balance of Nirmal Sagar, somehow, does not allow withdrawal of more than 12 MLD.

We can very easily see that it would be extremely desirable to bring Tata's Nano Factory from socio-economic perspectives to the City of Timbuktu. However, potable water supply from Nirmal Sagar will be a constraint. One alternative could be to treat the wastewater to the level of auto manufacturer's quality and quantity demand and implement some preventive environmental measures in the City wherein treated sewage water would be used and thereby saving potable water; which can be given to the 40,000 new population in the Tata's township. You should remember that Nano Factory needs 1 MLD of potable water (0.5 MLD for drinking, hospital and cafeteria and remaining 0.5 MLD for boiler) and 3 MLD industrial-grade water for manufacturing operations. Tata is willing to accept tertiary-treated City sewage provided the tertiary treatment meets the quality objectives of process water stipulated by them.

As a consultant you are required to submit written opinion on following questions.

(A) How will you satisfy the drinking water requirement of the City, drinking water requirement of Tata's township, process water requirement of Nano Factory and drinking water requirement of Nano Factory? In other words, show your water budget. You are allowed to use up to 60% of water demand of 135 Lit. per person per day using tertiary treated sewage recycled into a variety of operations. It is, however, important to specify applications and corresponding quantities that could be potentially implemented.

(B) How will you treat various streams of raw potable water and wastewater to achieve the desired quality and quantity of recycled water?

(C) Maruti Udyog would also like to setup an auto manufacturing facility in the City of Timbuktu, having pretty much the same requirements of quality and quantity of water. They are planning to launch "Squirrel" a tiny car which will be answer to Tata's Nano car. Can you accommodate Squirrel Maruti Udyog Factory in Timbuktu? Concentrate your discussion on the domestic and industrial water demands. Explain your answer through water budget.

9 CHAPTER

Municipal Solid Wastes Collection and Disposal

9.1 Overview of Municipal Solid Wastes from Urban Areas in India

9.2 Recommend Disposal Methods, Emissions: Lechates

9.3 The Old System of Dumping All Solid Wastes in Low-Lying Land

9.4 The Modern Sanitary Landfill Method

9.5 Solid Wastes Disposal by Other Methods

9.6 Bio-Methanation, Compost, Vermi-Compost for Small Set-ups

> *This is our last opportunity to make something useful and reusable out of waste.*

9.1 OVERVIEW OF MUNICIPAL SOLID WASTES FROM URBAN AREAS IN INDIA

Solid waste is the discarded and useless material generated in the various day to day functions of normal life. Materials are discarded because of various reasons such as not worth keeping, life of material is over, etc. They are inevitable in any society. Some of the solid wastes are broken glass, wrappers and covers, food waste, fruit and vegetable peels, papers, plastics from various products, wrappings, building debris, etc. The wastes are solid in phase.

The nature and quantity of solid waste generated depends upon the activity. In this chapter, only the solid waste that arises from urban areas and predominantly from residential and commercial establishments, also known as municipal solid waste (MSW) is discussed. Commercial establishments in municipal limits are market yards, malls, educational facilities, eateries, etc. Normally, MSW is non-hazardous in nature. Hazardous, infectious and toxic solid wastes are discussed separately in the next chapter.

9.1.1 Composition

The various components of the MSW are:

1. **Inorganic part**
   - Paper and card board
   - Glass
   - Metals (ferrous and non-ferrous)
   - Dust, dirt and sweepings
2. **Organic part**
   - Food waste
   - Vegetable and fruits peelings waste
   - Garden waste

Typical composition and properties of MSW are given in Table 9.1.

Table 9.1 *Typical composition and properties of Indian MSW*

Components	Composition (% By Weight)
Metal	0.2–2.5
Glass, ceramics	0.5–3.5
Food and kitchen waste	40–65
Paper	1–10
Textiles	1–5
Plastics/Rubber	1–5
Misc. Combustible	1–8
Misc. Incombustible	–
Inert	20–50
Typical Properties of Indian MSW	
Density (Kg/m^3)	250–500
Moisture content (% by Weight)	40–50
Waste generation (Kg/cap/day)	0.4–0.6

The composition of the MSW is divided into biodegradable and non-biodegradable fractions. The biodegradable (organic) fraction is complex in nature and is broken down into simpler inorganic and stable compounds by micro-organism (decomposers) as we have seen earlier in Chapter 1. The organic content may be 40 to 65 % of the total, the rest being inorganic. The inorganic non-biodegradable fraction is stable in nature.

The Cow Factor!

By comparison, the solid waste of European and American cities is more in quantity and has a higher content of paper, wrapping and packaging materials. These materials are combustible and, therefore, the calorific value of their waste is higher. In India, wandering cows and goats eat away vegetable and fruit peelings contained in the waste at street corners. Thus, Indian MSW has a much lower calorific value than western MSW.

9.1.2 Quantity

The solid waste generated per person varies depending upon the standard of living and other habits of the people. Hence, the quantity is different in different countries of the world. In Indian urban areas it is estimated to vary from 400 grams to 600 grams per person per day. The quantity of solid waste generated in some of the Western countries is more than double this quantity.

> ***Problem 9.1:*** Estimate the quantity of MSW generated per day in an Indian metropolitan city with population of 14 million (14×10^6). Assume per head solid waste generation to be 0.4 kg per person per day.
>
> Total quantity of solid waste generated
>
> $$= \text{Population} \times \text{Per capita waste generated}$$
> $$= 14 \times 10^6 \times 0.4 \text{ kg} = 5{,}600 \text{ T/day}$$

Construction and demolition (C&D) waste generated from an urban area depends upon the developmental activities in that area and is difficult to estimate. It could amount to 1000 to 2000 tonne per day for a large and bustling city like Mumbai. For this reason, the responsibility of managing the collection and transport of this waste is normally with the producer of the C&D waste. These wastes handlers are normally looking for low-lying lands to fill with their debris. Coastal areas, mangroves, wetlands and reserved lands are the likely places for such dumping – authorized or not.

Similarly, among the hazardous wastes generated from the urban areas are medical wastes from health care centers such as clinics, hospitals, research laboratories, etc., and special wastes such as electronic goods, car batteries, automobile tyres, discarded automobiles, etc. The management of such wastes is discussed separately. (see Chapter 10)

Collection

The solid waste generated in urban areas is collected at the source of generation (houses) in the small bins called as "dust bins" having capacity to collect 3 to 5 kilograms of waste. The waste is collected from door to door in a bigger bin of capacity 20 to 30 kilograms and brought to the outside of the building, from where the municipal or local authority collects the waste by transferring it to a special collection vehicle. The solid waste collection vehicle transfers this waste at a transfer station to either a bigger vehicle (generally with a compactor) or to a special train for onward journey to the disposal site.

The MSW is collected and carried from the transfer stations to the treatment/disposal facility with the help of special vehicles called "compactors". These special vehicles are equipped with arrangements for transfer of the waste from the waste containers. The waste is compacted to increase its carrying capacity. The MSW has a density of about 300 to 600 kg/M^3 without compaction and can be compacted to have density of 800 to 1200 kg/M^3. The compacters are vehicles which are completely covered and hygienic for the transport of the MSW.

A thought occurs as to why we should not operate all collection vehicles on renewable energy instead of petrol or diesel. This change would make an important contribution to saving import of fossil fuels. In the last World War all Mumbai's refuse carrying vehicles were operated on Producer gas in the absence of petrol/diesel. Today, we would not like Producer gas because of its emissions as it comes from coal. But, we have other alternatives.

> All transport vehicles can be made to operate on renewable source of energy instead of fossil fuels. To start with, a beginning should be made at least by using locally available CNG instead of imported diesel oil or petrol.

Segregation and Ragpicking

Ragpicking has been an activity in which the poorest of the poor have been engaged since time immemorial. They rummage through a filthy mess for whatever looks to them as salvageable and marketable. After their filthy work is done, there is an unbelievably efficient network of people engaged in transferring the salvaged material from the rag-pickers to the market. In response to suggestions over the years, the concept of "segregation of wastes" is catching on in India.

Segregation involves separating the organic wastes (kitchen and food wastes or "wet" wastes) from the inorganic wastes (rest of household or "dry" wastes) so that the rag-pickers are facilitated in their work and it becomes a little more hygienic for them to rummage through the wastes looking for salvageable material. The rag-pickers are fully in favour of improving their working condition.

The residents have to segregate the waste into two separate packages and the municipal transporter has to take the wet waste and the residue of the dry waste, in *separate* trucks to the disposal site so that the dry waste may be used for filling low-lying land while the wet waste may be used separately for composting.

9.2 RECOMMENDED DISPOSAL METHODS: EMISSIONS, LEACHATES

An urban area generates a variety of solid wastes daily and these wastes have to be taken away quickly and disposed of properly without damaging the environment or endangering public health. Urban areas have commercial, industrial and domestic activities all going on within them, each generating different types of wastes needing different handling arrangements. Logic requires that the objectives be first clearly outlined so that appropriate strategies for action may be developed.

9.2.1 Objectives

> Our basic objective should be to protect our health and the environment at least cost and with least use of our resources.

Hence, our operational objectives are as follows:

- The land requirement should be minimized.
- The waste should be converted into useful compost to the extent possible so as to benefit the soil and agricultural activity instead of just filling low-lying land. This way the life of the landfill site would also be increased.
- Pollution of the environment (air, soil, water, sub-soil water, etc.) should be avoided.

With these premises in mind, Fig. 9.1 shows diagrammatically how the wastes from different sources are recommended to be handled from any urban area:

- Hazardous wastes from selected industries and radioactive sources are to be handled separately as they may need to be immobilized first and then laid into a "secure" landfill for long-time isolation.
- Special wastes from hospitals, and E-wastes from selected sources should also be disposed separately after taking due precautions to protect health and environment.
- Construction and demolition waste should be used profitably to fill low-lying land.
- Highly organic food and such wastes from market places, hotels, industrial canteens, etc., may be taken separately to "waste-to-energy" or bio-methanation plants so as to benefit from their heat energy.

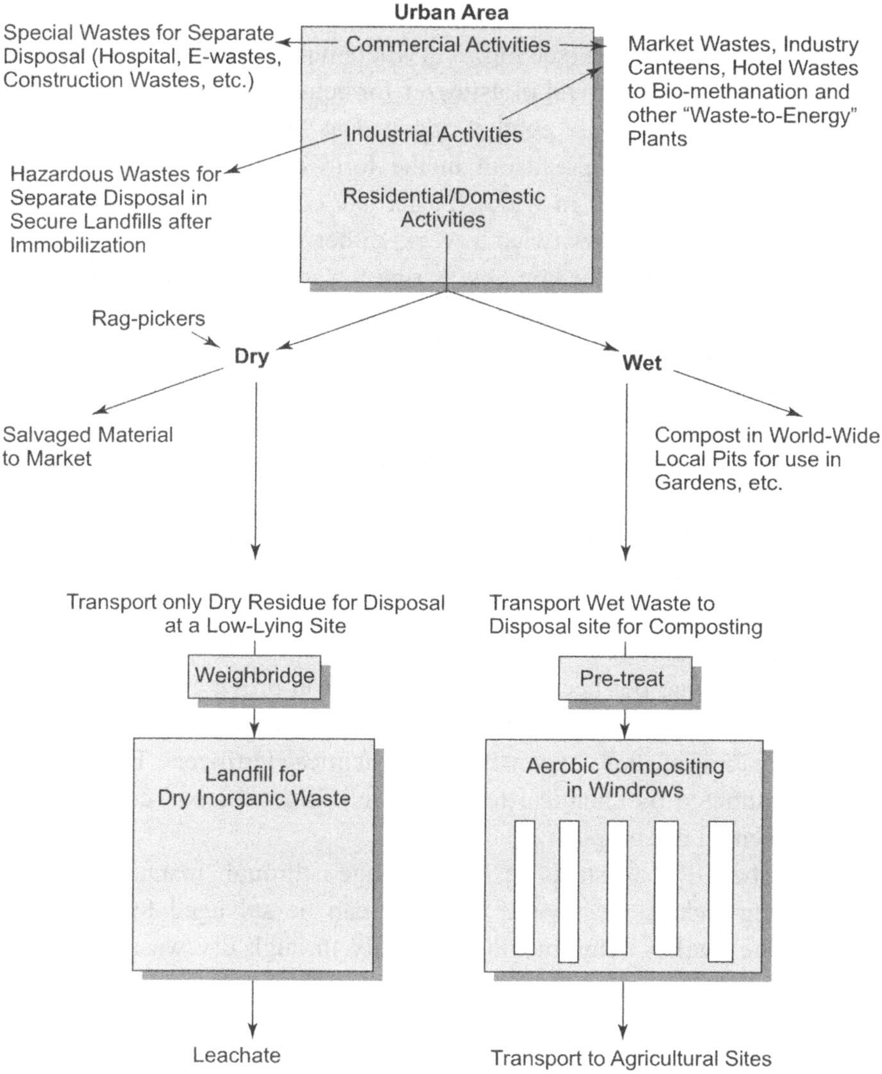

Fig. 9.1 *A master plan for solid waste disposal from an urban complex*

➢ The other solid waste (collectively known as municipal solid waste) from residential, domestic and other related sources be segregated into their "dry" and "wet" fractions for hygienic reasons and convenience of further handling by rag-pickers.

➢ A partial use be made of the WET waste to prepare compost, locally in local, ward-wise sites such as public and private gardens, parks, roadway strips, etc.

- ➢ The remaining WET waste be transported to the disposal site for making compost (see Fig. 9.1) where it may be subjected first to a sorter, shredder and moisturizer for being screened to 25–50 mm size and moisture content adjusted to 50–60%. The pre-treated waste is then spread out in the form of windrows (2.0–2.5 m high by 4.0–5.0 m wide) for aerobic composting. The heap has to be turned over twice a week, either manually or with special machinery till the compost is ripe for agricultural use 3–4 weeks later. Turning of windrows is often accompanied by release of offensive odours.
- ➢ The raw waste has a high Carbon: Nitrogen (C/N) ratio of about 25/1 to 35/1 which makes the waste unstable and decomposable. After composting the C: N ratio comes down to 20/1 to 15/1 when it is stable and usable in agriculture. It is called "manure" and improves the water holding capacity of the soil. If the waste is not sorted earlier, it makes its agricultural use as manure difficult.
- ➢ Once the compost is ready, it has to be transported for use to the agricultural fields. Agricultural use requires transportation @ 20–100 tonne per hectare per year. This can prove expensive. Often these fields may be distant adding to transport costs and thus making it more expensive than chemical fertilizers. Farmers are attracted by chemical fertilizers and forget the soil conditioning benefit of compost.
- ➢ The DRY waste is to be rummaged through manually by the rag-pickers to salvage whatever can be salvaged for resale in the market. (Rummaging manually through dry waste is not as unhygienic as rummaging through combined dry and wet waste).
- ➢ The residue of the dry waste after salvaging operation is complete, has to be transported to the landfill site, either for simple landfill or a sanitary landfill. Generally in India, simple landfill is done.
- ➢ Also, rainfall percolates through the dump-site generating some leachate which finds its way to a natural water course which gets polluted unless the leachate is treated separately. Leachates should be collected and treated before they are discharged to a nearby water course. Leachates are generally high in BOD and COD and dissolved organics. They are therefore, best treated in aerated lagoons before discharge. No such treatment is yet given to leachates in India. Care should be taken to ensure that surface run-off of rainwater is properly diverted into channels to flow

away from the disposal site, otherwise the volume for treatment would unduly increase during the rainy season.

Compost is a Soil-Conditioner

This is the method now required to be followed in India. No solid wastes disposal system makes profit for the local body. However, by composting we at least get a useful product out of the wastes and save on land which is a scarce resource. Also, by composting aerobically, we do not release any methane to the atmosphere.

9.3 THE OLD SYSTEM OF DUMPING ALL SOLID WASTES IN LOW-LYING LAND

The municipal solid waste generated in majority of Indian urban areas, however, have so far been handled without regard for the above principles. Firstly, segregation of waste was not enforced. The rag-pickers worked in filthy conditions. Secondly, the waste was a mixture of organic and inorganic wastes together, which went to the dumping ground for slow decomposition over many years, giving off methane and other gases to the atmosphere. The gases were not collected and used for their calorific value.

Disposal has today become an important matter for all its stakeholders, for the following reasons:

- Quantity of municipal solid waste generated is increasing.
- Land for solid wastes disposal is not available within feasible limits.
- Increasing amount of water, air and soil pollution occurs due to disposal.
- Stricter legislation for handling and disposal of MSW is enforced.
- Awareness and alertness of the city dwellers is also increasing.

Moreover, in the old system, the gases are given off to the atmosphere which according to present knowledge are considered harmful for global warming. Methane is considered to have a 25 times higher potential than carbon dioxide to cause warming. Thus its quantity has to be multiplied by 25 to get CO_2-equivalent. All these factors make the old system unattractive to use.

To reduce this emission, the sensible thing to do is to lay a network of pipes and manifold in the mass of solid waste dumped in the landfill to collect the gas and utilize it for a nearby industry or something. This would reduce the potential for global warming, and also bring some badly needed income to the local body.

Observation: Biogas released to the atmosphere from any old style solid waste disposal site located in India is approximately @ 0.42 g/person/day (because of low organic content of the waste). Anyway, with this rate of emission, biogas from say 1 million people would be 420 kg/day or 153.3 tonne per year. As Methane is approximately 50% volume of biogas, its emission = 210 kg/day. In terms of CO_2-equivalent, it is = $210 \times 25 = 5{,}250$ kg/day. Thus, total CO_2-equivalent emitted per day = $(210 + 5250) \times 365 = 1993$ tonne per year. This is a huge quantity going to the atmosphere from one dump!

Finally, when the disposal site is filled up, it must be "closed" using acceptable procedures and whatever biogas is still produced for some years should be collected and used, instead of allowing it to escape to the atmosphere. Only one dump so far in India (Gorai, Mumbai) has been given such attention (because it could get funds from the UN under their Carbon Credit scheme).

After the raddi-wala and the bhangar-wala have taken whatever they could, the inorganic fraction which is not recyclable (and which does not produce any gas) remains in the dumping ground for years without any further changes. The chemicals leach out slowly over a period through these dumps to the soils and sub-surface water near and far from the dumping ground get polluted by this leaching. Also the paper, plastics and other combustible or partially combustible waste-products start self-igniting through heat generated by microbial activity in the dump. The burning of these wastes emits various air pollutants, gases, odours and particulates in the atmosphere affecting the flora, fauna and human population in the vicinity.

9.4 THE MODERN SANITARY LANDFILL METHOD

Sanitary landfill is a more engineered way of disposing solid wastes but requires land, detailed planning, careful construction and efficient operation. Besides it is relatively more expensive. It is a method often used in western countries, generally in three different ways: sanitary landfill may either cover a depression, or a trench or an area. Sanitary landfill has generally been practiced using the full waste (both dry and wet fractions). In many countries it is still done using the full waste.

At the end of each day's operation, 150–300 mm layer of cover material (usually soil) is placed over the completed work. The covered material must be brought in by trucks or bull dozers from nearby land or borrow pit areas. The filling operation is usually carried out by spreading the MSW in thin layers and compacting. Each layer is compacted until the thickness of the compacted waste reaches a height varying from 2–3 m. The covering of each day's work is what makes the method "sanitary" as it covers the solid waste brought in during the day, thus excluding odours, fly and mosquito nuisance at the site.

However, as per Government of India's MSW (Management and Handling) Rules 2000, in India, composting of the wet fraction and Sanitary Landfill of the non-biodegradable, inert waste (the dry fraction), is required (see Fig. 9.1).

9.5 SOLID WASTES DISPOSAL BY OTHER METHODS

Incineration

It is the process of direct burning of waste in the presence of excess air (oxygen) at temperatures of about 800°C and above, liberating heat energy, inert gases and ash. In practice, about 65 to 80% of the energy content of the organic matter can be recovered as heat energy, which can be utilized either for direct thermal applications or for producing power via steam turbine generators. Incineration, however, is associated with polluting emissions which are of environmental concern, although in varying degree of severity, and may be controlled by installing suitable pollution control devices.

The Indian MSW contains high percentage of inert (40–65%) waste and relatively low percentage of putrescible organic matter. As mentioned earlier, the calorific value of Indian MSW is low, varying on dry weight basis between 800–1100 kcal/kg. Due to this self-sustaining combustion cannot be obtained from such waste and auxiliary fuels are normally required. Therefore, incineration becomes expensive and has not been preferred in India. Moreover, experience with incineration of hospital wastes indicates that incineration at lesser than optimum temperatures appears to lead to generation of harmful dioxins and smoke.

Refuse-derived Fuel (RDF)/Pelletisation

The process of conversion of garbage into fuel pellets involves primarily drying, separation of non-combustibles from garbage, size reduction and pelletisation after mixing with binder and or additives as required. The moisture content of the waste should not be more than 35–40% otherwise sun drying (1 to 2 days) is required, which needs more land. Earlier Refuse-derived fuels were used along with coal-fired boilers but now, due to stricter restrictions on emissions, specially designed boilers dedicated to RDF are required.

Incineration, pyrolysis and pelletisation options are all energy, laboratory, and land area intensive operations as well as having environmental pollution potential. Composting is a better method of treatment for organic fraction of MSW.

9.6 BIO-METHANATION, COMPOST, VERMI-COMPOST FOR SMALL SET-UPS

The objective of "Bio-methanation" is to create methane gas (which has fuel value) from the organic matter present in the waste. The process is also referred to as "**waste-to-energy**" because that is exactly what it does. Generally, wastes from market places, industrial canteens, kitchens, hotels, etc., are rich in organic content (waste food, vegetable peels, etc.) and therefore very suited for separation at source and treatment by bio-methanation or gasification or some such process to convert waste to energy.

Bio-methanation or Waste-to-energy Plants

Mention must be made of two types of "waste-to-energy" plants which are often provided, one at village level and another at industry level to convert waste matter rich in organics into useful energy-giving fuel.

- Gobar gas plants in rural areas
- Waste-to-energy plants in specific locations like market places, industrial canteens, etc.

Gobar gas plants They are the simplest in this category and are generally small in size, suitable for individual families in rural areas using human excreta and/or manure from a few cows or buffaloes.

For agriculture and milking purposes, rural families tend to maintain a small number of say 3 or 4 cows or buffaloes each. Their dung helps run a bio-gas plant and produces bio-gas which helps the family to cook and heat as required without depending on their womenfolk to go in the adjoining fields and forests to collect wood/twigs to light fires at home for cooking or heating. This, incidentally, reduces deforestation and saves the womenfolk their time. Gobar gas plants also prevent indoor air pollution, encourage better sanitation in rural areas by ensuring that toilets are constructed and feces is drained to the bio-gas plants rather than left in the open.

Observation: All such activities (Gobar gas plants along with rainwater harvesting) when carried out on a large enough scale can change the face of rural India, putting more money and more time in the hands of rural families.

Waste-to-energy plants CII-Godrej publication indicates that a 4.5 MW biomass based power plant installed in Karnataka consumes about 150 tonne of biomass per day to generate about 1 lakh units of electric power. The problem of fuel availability afflicts power generation from agricultural and animal wastes. About 100 tonne/day of cattle dung is required to generate 300 kW of power. Such large amounts of dung would be only available from large-scale cattle sheds. Thus, the full potential for power generation from agro-wastes and such other sources yet remains relatively unexplored in India.

However, several activities produce organic wastes in reasonably large quantities at their sites such as at market places, industrial canteens, hotels, etc. One could locate suitable sites, on case by case basis, for installing individual *"waste-to-energy"* plants for which technology is available. For example, at market places where much solid and liquid wastes high in organic content are collected, waste-to-energy plants can come up to advantage. These plants would convert the organic matter into stable end-products while releasing the nutrients for recycle. One could also get Indian or foreign-made plants on B.O.T. basis so that least capital investment is needed from the local body (see Section 7.5). The local body will only have to give land on lease basis and ensure maintenance will be adequate.

Thermax, Pune is reported to be tying up with Lambion to provide energy generation plants from wastes in India. The example of a bio-methanation plant recently set-up in Chennai for a market waste of 30 tonne per day can also be quoted.

Mention must be made of the *Bio-methanation plants (Nisargruna)* developed by the Bhabha Atomic Research Centre (BARC) which are being successfully run in India for treating food and kitchen wastes from large canteens of industries and offices, camps and religious fairs, food processing establishments, market places, etc. Wastes that cannot be treated by bio-methanation are coconut shells, egg shells, big bones, plastic/polythene, glass, metal, sand, slit, debris and building materials, wood, cloth/clothes, ropes, nylon threads, batteries, tyres/rubber, hazardous and chemical industries waste, etc., and are excluded.

Composting for Small, Residential Set-ups

As we saw earlier, composting is a process for stabilizing the solid waste aerobically in above ground windrows from a high C/N ratio of 35/1 to a lower ratio of about 15/1. It needs some turning over of the windrows about twice a week to bring in some fresh air and it takes about 3 to 4 weeks to reach a stable condition. The stable humus becomes a good soil conditioner. Compost also provides trace nutrients and small quantities of nitrates and phosphates good for plant growth.

Composting can also be done anaerobically, especially for small residential set-ups to avoid turning over of the material for aeration. Anaerobic composting takes more time than the windrow method, and produces some odour when the pit is opened. Yet, it is the simplest method to produce compost on a small scale. Just place the waste in a ditch or pit and leave it for decomposition to take place. It is not favoured for larger volumes.

For successful composting on a small scale, segregation at household level is very important. After segregation, a family of 5 persons may produce about ½ kg of wet waste daily. This can be easily composted in a large earthen pot, watered daily with a glass of water. Each day's wet garbage is covered over and decomposes as new layers are added. The waste can also be vermi-cultured, if desired, especially if a group of families is concerned.

Residential set-ups produce solid wastes daily whose disposal has to be so engineered that it can give some compost for use or sale locally by the family concerned. Composting pits of size 1.6 m × 1 m × 0.6 m are set-up for processing biodegradable waste. On an average, for a housing complex, a set of three such pits is required for 100 families.

All biodegradable waste generated in parks and gardens (including dry leaves) can also be composted. Addition of water and cowdung or kitchen waste can speed up the process.

Vermi-composting of Kitchen Wastes

Vermi-composting seems to be preferred over plain composting for small set-ups. Vermi-compost is the droppings/castings of the earthworm after intestinal digestion of organic matter. These castings are high in nutritive value and high-grade organic fertilizer. This helps in growth and multiplication of beneficial micro-organism by creating favourable environmental conditions. In vermi-composting organic waste is broken down and fragmented by earthworms resulting in a stable non-toxic material with good humus that can be used as a soil conditioner. There are over 4000 species of earthworms identified till date. Common Species used for vermi-composting are *Eisonia foetida* (*E-foetida*), *Eudrelus eugeniae* and *Perionyx excavatus*.

Vermi pits are generally made having a maximum depth of about 0.6 m, width of 1.5 to 2 m and can have any suitable length. Area required is approx. 1.0 m^2 per family of 5 persons. In the pit, the bottommost layer of 10 cm thickness consists of baggase, coconut coir or saw dust. On this, a second layer of 10 cm thickness of decomposed cow dung is spread. The mass is inoculated with earthworms at the rate of 4000–5000 nos. or 1 kg per sq m area. Above this partially decomposed organic matter is filled up to top of pit. The pit is covered with gunny bag or rice husk. Water is sprinkled on it daily to maintain the moisture.

The earthworms "eat" the organic waste and vermi-castings start getting deposited on the surface of the heap. The harvesting of vermi-casting starts from the sixth week from receipt of garbage. The vermi-castings collected are sieved by using simple sieves. The following parameters/conditions are important and need to be maintained for proper working of vermi-composting pits:

- ➢ pH should be within range of 6.8–7.5.
- ➢ Optimum temperature required for vermi-composting is 25–35 °C.
- ➢ Moisture of about 30–40% is required for growth and reproduction of earthworms.
- ➢ Earthworms are very sensitive to heavy metals and toxic substances or pesticides.

Table 9.2 *Typical compost properties*

Compost Properties	
pH	6.5–8.0
Moisture content	25 %
Nitrogen	1–2 %
Potash	0.2–1.0 %
Phosphorus	0.4–1.0 %

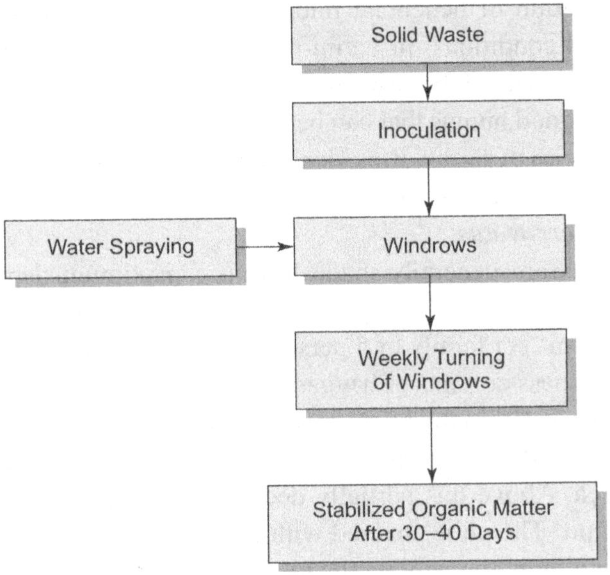

Fig. 9.2 *Steps in composting with aerobic windrows*

As We Said Earlier

By composting we at least get a useful product out of the wastes and save on land which is a scarce resource. Also, by composting aerobically, we do not release any methane to the atmosphere. Segregation of waste makes the work of rag-pickers more hygienic.

EXERCISES QUESTIONS

1. Explain any type of "waste-to-energy" plant you know.
2. Why is "segregation" desirable to have? When can segregation end up as a failure?
3. When and where would you recommend the old anaerobic composting method (Nagpur method)?

PROJECT/ASSIGNMENT TOPICS

Briefly discuss each of the following options for implementation, marking those which you think are feasible to adopt, and those which might possibly have any undesirable consequences.

Soft Options

- Hold seminars and discussions on solid waste collection and disposal methods.
- Adopt segregation of wastes at source.
- Adopt traditional system of composting. Transport compost to agricultural sites out of town.

Hard Options

- Conserve use of fossil fuels for transport of solid wastes and transport of compost.
- Promote use of landfills with biogas collection and usage.
- Follow Supreme Court's recommended practice fully and faithfully.
- Follow recommended practice fully + separate disposal of hospital and E-wastes.

10 CHAPTER

Collection, Treatment and Disposal of Toxic and Hazardous Wastes

10.1 Overview of Collection, Treatment and Disposal of Hazardous Wastes

10.2 Treatment, Storage and Disposal Facilities (TSDF)

10.3 Process Changes Incorporating 4-R's Useful in Handling HW

10.4 Remediation Technologies for Contaminated Sites

10.5 Collection and Disposal of Hospital Wastes

10.6 E-wastes and Electrical Equipment Wastes

We either give an expensive "burial" to hazardous waste or "cremate" it, in the hope that it will rest in peace forever!

10.1 OVERVIEW OF COLLECTION, TREATMENT AND DISPOSAL OF HAZARDOUS WASTES

At the outset, we must distinguish between municipal solid wastes which we saw in the previous chapter and hazardous wastes discussed here. Municipal solid wastes mainly come from domestic sources, offices and shops, whereas hazardous wastes come mainly from industries. Hazardous wastes are defined as "hazardous" because of any one of the following three characteristics:

- Properties, or
- Source, or
- Concentration

They may have one or more of the following hazardous properties such as toxic, ignitable, corrosive or reactive with water, or they may be generated in certain specified processes which are known to be hazardous sources (irrespective of their concentrations) or they may be specific products in solid or semi-solid form (such as sludges) in excess of certain concentration limits specified by the Central Pollution Control Board (CPCB) or the State Pollution Control Boards (SPCBs) in India.

Rapid industrial and economic growth in India in the recent past generated a large variety and quantity of hazardous wastes that continue to contaminate soil and water and negatively impact human health. Such wastes are usually by-products of industrial operations and include:

- Acids and alkalis.
- Heavy metals such as arsenic, cadmium, chromium, lead, and mercury.
- Oily wastes and tarry residues.
- Processes that utilize different categories of oil, petrochemicals, and solvents.
- Products such as polyvinyl chloride (PVC).
- Chemicals such as polychlorinated biphenyls (PCBs).
- Unused pesticides from containers.
- Waste products from photocopiers; electrical and electronic equipment.

- ➢ Pathogenic and other contaminated material from hospital wastes including by-products from incineration such as dioxins and furfurans recognized as highly toxic materials affecting all forms of life.
- ➢ Persistent organic pollutants (POPs) and other toxic substances generated from new industrial processes that cannot be dealt with by currently available technologies and whose ill effects have begun to far outweigh their estimated benefits.
- ➢ Red muds from bauxite processing, etc.

Until late nineties, the CPCB or SPCBs did not have any comprehensive management plans or institutional capacities to address the impacts of hazardous wastes, either by reducing their production or their risks through control technologies and other mitigation measures or containing them safely in proper disposal facilities such as designed/engineered landfills. We had a set of dysfunctional rules to regulate them, however.

Generally, industries in India used to do one of 3 things (for which reason "due diligence" studies were generally recommended to be carried out before purchase of land):

- ➢ Store hazardous wastes (HW) on their premises (in improperly designed facilities, e.g., Zuari Chemicals in Goa), or
- ➢ Dispose of them illegally in low lying areas in open pits or in makeshift "dump sites", or
- ➢ Illegally dump on public or abandoned lands, rivers and oceans (e.g., H-acid case of Bichhri village in the Udaipur District of Rajasthan).

In addition to hazardous substances generated due to industrial processes used within the country, there was additional impact of hazardous wastes brought from other countries either for genuine recycling or brought in illegally for final disposal.

To regulate hazardous waste generation and disposal, the Government of India (GoI) promulgated the *Hazardous Wastes (Management and Handling) Rules* in 1989 (amended in 2000 and 2003); the *Biomedical Wastes (Management and Handling) Rules* in 1998 (amended in 2003); and the *Batteries (Management and Handling) Rule* in 2001. Regarding chemical safety, the GoI issued the *Manufacture, Storage and Import of Hazardous Chemicals Rule* in 1989 and *Chemical Accidents (Emergency Planning, Preparedness and Response) Rule*, 1996. Other related Acts include the *Amendments to Customs Act, 1962*, *Amendment to Major*

Ports Act, 1962 and Amendments to Foreign Trade (Development and Regulation) Act, 1992. Furthermore, under the Basel Convention (1992) the member countries committed themselves to minimize the trans-boundary movement of HW and to prohibit exports to countries that have banned such exports. However, most of these rules are yet to be effectively implemented.

An international organization, *Greenpeace* first drew attention to the import of hazardous waste to India in violation of the Basel Convention. Another non-governmental organization (NGO), the Research Foundation for Science, Technology and Natural Resource Policy, filed a writ petition before the Indian Supreme Court seeking its intervention for effective management of hazardous waste. This led the Indian Supreme Court to establish a High Power Committee (HPC). In addition to monitoring dumping and import of hazardous waste, the HPC has investigated the systemic weaknesses that are preventing effective remedial actions.

The MoEF has now brought amendments to the existing Rules under which waste streams are defined as "hazardous" based on either their properties, or source or concentration. This will facilitate improved monitoring and compliance. Furthermore, the MoEF has proposed a list of processes generating HW and specific definitions of some additional terms that would improve understanding of the Rules and make identification of HW streams easier.

10.2 TREATMENT, STORAGE AND DISPOSAL FACILITIES (TSDF)

The fact remains that the hazardous material one is dealing with has little or no resale value, and the possibilities for treating HW are, indeed, few (and expensive) such as the following:

- Long-term storage in "secure landfills" using solidification, chemical fixation or encapsulation to stabilize the waste prior to long-term storage.
- Detoxify through chemical transformation or destroy through oxidation or reduction.
- Destroy through biological transformation (aerobic or anaerobic).
- Incinerate to reduce volume/weight before burial or storage of residue.

TSDF facilities are generally located at carefully selected places with regard to geological and other factors (discussed below) and strategically located to be near Common Effluent Treatment Plants from where much material may come falling under the HW category.

10.2.1 Typical HW Disposal Layout

Figure 10.1 shows a typical layout of a HW disposal facility. It has separate receiving centres for:

- ➢ Hazardous wastes of a general nature,
- ➢ Hazardous wastes to be incinerated (e.g., spent solvents), and
- ➢ Hazardous wastes to be stabilized before placing in the secure landfill.

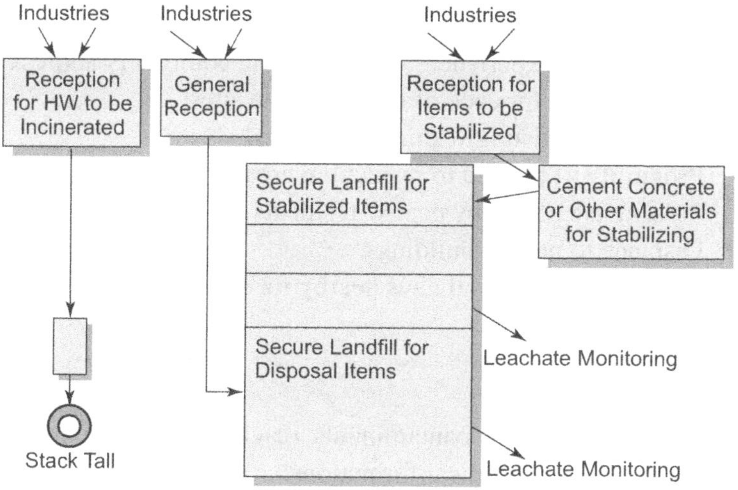

Fig. 10.1 *A hazardous waste disposal site*

The stabilization method used depends on the type of waste to be stored. Stabilization reagents generally used are Lime, Cement, Fly Ash, Clay, Gypsum, etc., *solidification* is done for oily sludges, and effluent treatment plant sludges, while *chemical fixation* is done for metal and other plating sludges and spent catalysts, and *encapsulation* is done in cement concrete for bottom residues and in glass for radioactive wastes. Costs vary from ₹ 2,500–5,000 per MT of waste. Hence, every effort is made to reduce the quantity to be stored.

Incineration facility may be located either at the same site where the secure landfill is set up or at a different site since its location may be influenced by wind direction and other factors (see Chapter 5).

> As disposal of hazardous material is an expensive proposition, it stands to reason that every effort should first be made to reduce its quantity as far as possible. Cleaner technologies have to be used.

10.2.2 Identification of Sites for Secure Landfills

The essential objectives in determining a suitable site are:

- Minimizing risk to human health,
- Minimizing environmental impacts, and
- Minimizing financial costs (without sacrificing the previous two objectives).

One major challenge faced by many waste managers in India relates to locating suitable sites for hazardous waste disposal. Site selection criteria for waste disposal have been developed over the years from a large number of studies and actual experiences. These criteria commonly address issues such as the following for each of the candidate sites:

- Population existing within 500 metres.
- Proximity of the site to residential areas.
- Present use of site by nearby residents.
- Distance to nearby buildings.
- Presence of roads/railways nearby for access.
- Land use/zoning.
- Ownership.
- Growth potential.
- The likelihood of contaminants reaching a source of drinking water due to geological formations.
- Depth of ground water.
- Likely impacts of flood, and earthquake.
- Air pollution potential.
- Any critical environmental features.

However, while addressing similar issues, the nature of their classification and presentation varies widely. One of the strengths of the Hazardous Waste (Management and Handling) Amendment Rules, 2000, is the detailed protocol outlined for notification of disposal sites. The key steps are Site Selection and Environmental Impact Assessment (EIA). The framework for Site Selection process needs an illustrative approach, where GIS mapping would be helpful.

Siting a landfill or other facility requires consideration of numerous technical criteria, socio-economic factors, and applicable laws and regulations. Massive amounts of spatial data have to be processed for waste facility siting. Manual analysis of such data is time consuming and often tedious. Furthermore, when there are two or more sites available, there is need to consider a wide range of factors.

> Using satellite imageries and suitable software package, it is possible to locate lands for waste disposal that are away from habitation and risk areas and not earmarked for agriculture, forestry or other land uses.

Site selection is a screening process that uses the available information to identify a preferred location for a solid waste landfill. The first screening criterion includes exclusion of critical areas identified based on distances recommended by experts. This initial screening will result in a number of candidate sites being eliminated. To narrow the numbers further down, a second screening criterion can be used. Through the second screening criterion, specific sites can be short-listed. These short-listed areas can be reserved for developing landfill sites and can be incorporated into the State development plan as "reserved" areas. During the initial screening of the selected region, the first step is the identification of critical data elements based on regulations and guidelines with their regulatory buffering distances. On the basis of identified elements, data collection can begin.

Waste disposal sites for hazardous wastes have always to be "secure landfill sites" as no chances can be taken with leachates from them polluting the environment. In fact, no leachates must occur and seep into the ground. For this reason waterproof membranes must be used to line the bottoms of these landfills to make them secure. Again, no chances can be taken with membranes. So we must have two or three membranes, one below the other, to be sure. All monsoon run-off must be diverted away from the landfill by use of catchwater channels. The hazardous waste to be placed in it must also be "stabilized" in cement concrete or similar material to prevent its possible dissolution in water.

10.2.3 Some Requirements under HW Rules

Pursuant to the Hazardous Wastes (Management and Handling) Rules, 1989 the GoI requires plans for the following in order to make sure that the proposal has been fully thought out for its safety aspects:

- Leachate management
- Groundwater monitoring

- Fugitive emissions
- On-site/off-site emergency response procedures
- Landfill conditioning and post-closure activities

Leachate management and groundwater monitoring activities have been conducted by the concerned agencies with varying degrees of robustness. However, ground water monitoring plans need to be studied in depth and action plan for identifying fugitive emissions is to be developed. The April 2008 accident in Gujarat at the Bharuch Enviro Infrastructure Plant reinforced the need for more definite guidelines for on-site and off-site emergency response mechanisms. Also, post-closure measures need to be executed for the abandoned landfill sites, but this also requires significant financial investments. In the absence of adequate funding, these sites cannot be used for any useful purpose.

10.2.4 Monitoring Facilities for Hazardous Wastes

The ability of SPCBs to effectively control the HW generating units will depend on the availability of appropriate equipment for monitoring as well as analysis of hazardous wastes and also on the routine of collection of samples from the source of generation as well as periodic and surprise inspections of hazardous wastes treatment, storage and disposal facilities. The SPCB's authority will largely diminish if it fails to effectively undertake periodic and unannounced compliance monitoring site visits. Therefore, provision of adequate laboratory facilities in each SPCB is necessary.

10.3 PROCESS CHANGES INCORPORATING 4-R's USEFUL IN HANDLING HW

As the standards for disposal of hazardous wastes are becoming more stringent over time, industries have accepted the fact that "Compliance along with Reduction, Recovery, Recycle and Reuse (4 R's) is cheaper, rather profitable, than non-compliance". Industries have overwhelmingly responded to the "Reduction at Source" principle by doing effective process modifications. Most large industries have already incorporated the principle of 4 R's as integral part of routine process activities. Several examples of such efforts are available. Due to these efforts which are more than what are required for compliance, most industries are exploring the possibility of taking Clean Development Mechanism (CDM) benefits. For example Birla Corporation Ltd., Satna, has taken CDM benefits by recovery and reuse of waste heat for power generation and energy conservation activities.

Cement manufactures have taken actions for incorporating the 4 R's into their production process. Wastes generated in other industries such as pharmaceuticals have been successfully utilized by cement industry for various purposes ranging from additives in cement to fuel in incinerators for captive power generation. Plastic waste has been reprocessed by ACC Kymore, M.P. for heat recovery.

Many pharmaceutical industries have indigenously developed chemical recovery plants for reducing and reusing the hazardous waste generated. Industries have started developing symbiotic relationships among each other for effective recycling of hazardous wastes. For example, Iron sludge generated from Paracetamol manufacturing units is used as a filler material in the cement industry.

Low manufacturing cost implied by less energy use along with increased demand offer attractive solution for recycling of hazardous metal wastes (ferrous and non-ferrous). For example, large chunk of waste recovered from ship breaking industry in Alang, Gujarat is not only recycled successfully and reused but also supplements primary production of such products by providing recycled material at much lower costs and at much lesser carbon foot print. Furthermore, global warming potential of such recycling processes is much lower than that of conventional manufacturing processes for virgin products.

More examples can be given of recovery and reuse of hazardous wastes successfully done in India. Special drives should be conducted for identifying yet more recycling opportunities and developing separate collection mechanisms for such hazardous wastes. Techno-economic feasibility of various treatment options for such waste should be validated at pilot level before full implementation.

10.4 REMEDIATION TECHNOLOGIES FOR CONTAMINATED SITES

As stated in Chapter 4, it is always advisable to carry out a "due diligence" survey before land is purchased for setting up a new factory or any other project. Only a "due diligence" survey would reveal how the previous owner had used his land. After all some 20 or 30 years ago, nobody took HW seriously and even Pollution Control Boards often specified that HWs be disposed of in the factory's own premises till further orders! This being the case, one could get a rude shock after the land is purchased. The site remediation work has then to be done at the cost of the new buyer.

A variety of remediation technologies are available such as Bio-remediation, Phytoremediation, Plain stabilization, etc., depending upon the time and funds available for implementing them. Education and training on remediation technologies and their associated advantages and disadvantages, appropriate use, costs and investments, risk factors, and applicability to differing scenarios can be an ongoing effort for pollution control authorities. Technologies are continually updated, chemical and contaminant risks are routinely reviewed, revised, or challenged, and best engineering practices and guidelines are frequently documented and shared, within India and internationally.

Over twenty years ago, the US EPA developed and still continues to support the Hazardous Waste Clean-Up Information (CLU-IN) outreach initiative to provide information about innovative treatment and site characterization technologies to the hazardous waste remediation community. Beginning in 1987 with bulletin messages, newsletters, announcements about training and workshops, and meetings of interested stakeholder, CLU-IN has grown to include in 1996 one of the most comprehensive remediation websites available and has become an important forum for remediation stakeholders. CLU-IN website provides information on tools for federal and state personnel, consulting engineers, technology developers and vendors, remediation contractors, researchers, community groups, and individual citizens. Access to the remediation technology information on the CLU-IN website is available to everyone, Till now there is no such initiative in India.

10.5 COLLECTION AND DISPOSAL OF HOSPITAL WASTES

Hospital wastes are classified as "hazardous wastes" and regulations have been laid down for their careful handling and disposal. The wastes emanate from hospitals, clinics, nursing homes, laboratories, blood banks, etc. Each State's Pollution Control Board is the responsible control agency in India. The topic of medical wastes has been selected for inclusion in this book because it is one of the most rampant and widespread health problems in Indian cities and affects the common man.

Hospital wastes can affect the health of the nursing and ancillary staff in the hospitals as well as of the rag-pickers and others outside the hospitals and clinics. In fact, in a country like India where a large number of people are poor and illiterate, a thriving "business" can go on in unauthorized reuse of needles and sharps and expired drugs leading to a chain of persons with vested interests waiting to take advantage of any laxity in care or control.

10.5.1 Sources of Wastes

The first seminar on the subject was held in 1997 in Mumbai at the joint initiative of Tata Memorial Hospital, Mumbai and an NGO called "Bombay First".

As very little Indian data existed at that time, US data, given in Table 10.1, was used to indicate typical sources of waste.

Table 10.1 *US data on sources of waste*

Sources	% in US (1990)
Hospitals	77.0
Nursing homes	6.4
Physicians offices	5.7
Clinics	3.6
Laboratories	3.3
Dentists offices	1.6
Blood banks, veterinarians	2.4

10.5.2 Typical Problems Caused

Typical health and environmental problems caused by inadequate handling of medical wastes in India are listed below:

Health/Infections

- Infections spread through contaminated articles mainly to hospital staff and waste handlers.
- Health problems caused to general public through resale of infected needles, etc., through unscrupulous rag-pickers and their network.

Odour nuisance If wastes are not removed promptly in hot climates, odours build up.

Offensive to sight If wastes are discharged untreated with other municipal wastes in open wayside bins, they present an offensive sight. Animals tend to make it intolerable sight.

Emissions and ash from incineration Incineration of PVC-containing plastics at inadequate temperatures can lead to fumes of dioxin and furfurals, etc. Sharps and needles contained in the waste show up in the incinerator ash.

10.5.3 Environmental Disease Potential

- Hepatitis B and AIDS/HIV from blood contaminated syringes/needles. According to WHO, this problem is mainly with hospital staff (10–20 cases/1000 workers).
- Drug addicts may use discarded syringes/needles.
- Unfortunately, some uneducated people in India believe that every cure requires an injection. Some unscrupulous persons take advantage of this belief to use discarded needles for injections putting the patient at risk.
- Scavengers/rag-pickers and refuse workers are at much risk from infected wastes, sharps and needles (180 cases/1000 workers according to WHO).

HBV infection more likely than HIV as it exists longer.

10.5.4 Current Scenario in India

- Lack of awareness all round, as yet.
- Very little data on environmental health effects of wrong disposal practices.
- Inertia (KalHoga, Main hoonna!).
- Poverty.
- The pressure of rag-pickers and recyclers is high because of resale value.
- Thanks to the efforts of Pollution Control Boards, hospitals are beginning to progress from indiscriminate disposal to controlled disposal. Small establishments, however, are doing very little for controlled disposal. Wider coverage is awaited.

10.5.5 Planning a Hospital Waste Management Program

Average Quantity of Hospital Wastes

As is true with all other forms of wastes, the wastes from Indian hospitals are lesser than wastes from hospitals in western countries as shown Table 10.2.

Table 10.2 *Waste produced per bed per day in some countries*

Country	Hospital Waste (kg/day/bed)
USA	4.5 kg/day/bed
Latin America	2.6–3.8 kg/day/bed
Netherlands, France	2.5–2.7 kg/day/bed
India	1.13 average total [0.52 (infectious)]

Waste handling in hospitals needs separation of wastes and color-coded bags for infectious and *non-infectious* portions and for different types of disposal. This has to be planned and made known to every waste handler. Periodic surprise checks are also needed.

Needles and sharps need destruction before disposal as they have resale value. Destruction is important before disposal. Destruction is the most important step for hospitals and even for small clinics, nursing homes, etc., in a city. Do not send these items to incinerators (otherwise ash handling becomes a problem). Do not solve one problem and create another

Infectious waste: Use of "No-burn" technologies such as auto-claves, hydro-claves, etc., followed by shredders, is preferred since burning at low temperatures may produce dioxins and furfurals if PVC-containing wastes are not removed from them earlier.

Body parts need separate disposal directly to incinerators or crematoriums.

Special wastes (Radioactive wastes, Cytotoxic drugs) need separate disposal as per separate guidelines.

Only ordinary, non-infectious waste can go directly to municipal collection system.

Note on incinerators Only a few, centralized, well-operated, incinerators should be preferred in any city for handling body parts, organs, infected tissues and cytotoxic drugs. The following items must be excluded from incinerators:

- Sharps/needles must be excluded.
- PVC-containing plastics give toxic dioxins and furfurals if burnt at lower temperature than required (>400°C) as it happens at start and end of a cycle. (Use of LDPE, HDPE, Polypropylene, no problem).

- Also, Hg and heavy metals dangerous in emissions. Chimneys are required to be min 30 ft. high.

10.5.6 Role of State Pollution Control Boards

The State Pollution Control Board is required to be a strong Regulator. Therefore, it cannot be an Operator. A few privately managed collection, treatment and disposal companies are urgently needed. A beginning has already been made. Competition is also the best way to reduce prices for the service.

The medical profession would like to off-load the work of waste disposal to anyone willing to take responsibility. It is hoped some persons with entrepreneurial spirit will see the "business opportunities" involved and plunge in. New MoEF specifications in this respect are welcome and must be followed in all respects.

10.6 E-WASTES AND ELECTRICAL EQUIPMENT WASTES

The 20th century's technological revolution, brought upon the consumer, a demand for constant upgrades to newer, faster, and more advanced models of appliances for offices and homes. This phenomenal demand has been most radically observed in the computers and mobile phones industry. The computer industry continues to introduce new technological advances and "upgrades" every 18 months on an average. In case of mobile phones, the obsolescence is even more rapid. The intense rate of wasting out of devices has, in turn, resulted in increasing quantities of electronics being added to the waste stream. This particular waste is composed of electrical and electronic components, *viz.*, entertainment electronic systems, telecommunication equipment, electronic tools, etc.

"E-waste" is used as a generic term embracing various types of waste containing, principally electronic components. Electronic waste contains both valuable as well as harmful materials and hence these materials require special handling and management. European Union's Waste Electrical and Electronic Equipment (WEEE) Directive lists the under mentioned products, at end of their useful life, as E-waste (EU, 2000).

- **Large Household Appliances**: Refrigerators, dryers, washing machines, air-conditioners, etc.
- **Small Household Appliances**: Vacuum cleaners, irons, toasters, coffee machines, etc.

- **IT and Telecommunication Equipment (ITAC)**: PCs, Laptops, Mobiles, Telephones, Fax Machines, Copiers, Printers, etc.
- **Entertainment and Consumer Electronics:** Televisions, VCR/ DVD/CD players, Hi-Fi sets, radios, etc.
- **Lighting Equipment**: Fluorescent tubes, sodium lamps, etc. (Except bulbs and Halogen bulbs)
- **Electrical and Electronic Tools:** Drills, electric saws, sewing machines, lawn mowers, etc. (Except large stationary tools/ machines)
- **Toys, Leisure, Sports and Recreational Equipment**: Coin slot machines, treadmills, electric train sets, etc.
- **Medical Instruments and Equipment:** CT Scanners, X-ray machine, etc.
- **Different types of Monitoring and Control Equipment**
- **Automatic Dispensers:** Bank ATMs, issuing machines, etc.

Estimated fraction of the major categories is given in Fig. 10.2.

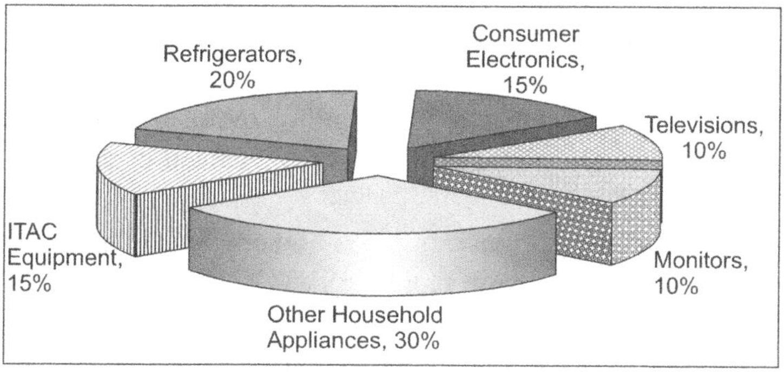

Fig. 10.2 *Types of E-waste (Source: UNEP, 2005)*

Principally, all equipment which needs electricity to work properly is either electrical or electronic. Each electrical or electronic product consists of a combination of several basic building blocks. The basic building blocks are (EPA, 2004):

- Printed Circuit Boards/Assemblies (PCB)
- Plastics containing flame retardants
- Cables, cords, wires and connectors
- Mercury switches and breakers
- Display equipment, such as cathode ray tubes (CRT) and liquid crystal displays (LCD)
- Sensors and Accumulators

- Data storage media
- Light generating devices
- Capacitors, resistors and relays
- Batteries

Handling and management of E-wastes has become a problem demanding special attention on account of the two salient characteristics mentioned hereunder.

10.6.1 Growth-trends in Volume of Discarded Electronic Devices

Due to the extremely rapid rates of obsolescence, electronic goods produce much higher volumes of waste in comparison to other consumer goods. The increasingly accelerated evolution of technology combined with rapid product obsolescence has effectively rendered everything dispensable. A user now only occasionally takes a broken product to a repair shop, as replacement is often cheaper and more convenient than repair. The average lifespan of a computer has shrunk from four or five years in 1997 to two years in 2004 (SVTC-2004). Similarly, mobile handsets are turned over every 10–12 months.

It has been estimated that 20 to 50 million tonne of waste originating from electrical and electronic equipment is generated every year, in the world.

E-waste constitutes as much as 2–5% of the US municipal solid waste stream and continues to grow rapidly (SVTC–2002). In 1998, E-waste constituted 4% of the municipal waste stream. Its volume is expected to increase by at least 3–5% per annum, a rate nearly three times faster than the average growth rate of municipal waste streams (UNEP, 2005).

Generation of E-waste worldwide varies considerably from country to country. While Thailand reports only 60,000 Mt/yr in 2003, Germany reports over 1 million MT/yr in 2005 and the USA reports over 2 million Mt/yr in the year 2000. There is a wide disparity in the data for E-waste generation in India. Different sources report different numbers ranging from 50,000 to 100,000 Mt/yr in 2005.

India is one of the leading consumers in the world of white goods including televisions and other electronics (Fig. 10.3). It is presently the 2nd fastest growing computer market in the world after China. While sale growth in the western economies has stagnated in % terms (though still considerable waste is generated due to old computers and peripherals being discarded) India is seeing exponential growth.

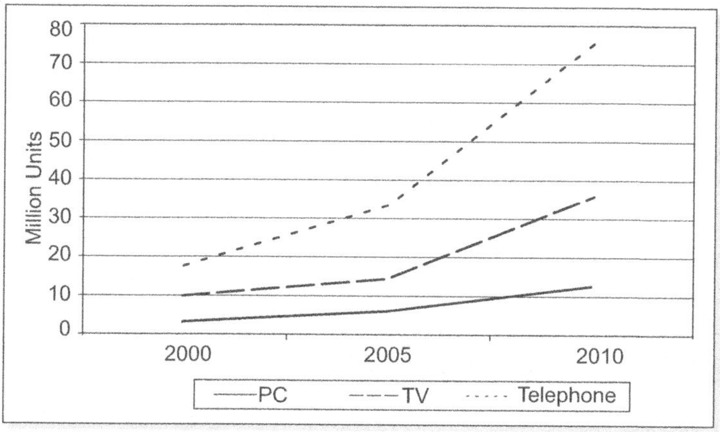

Fig. 10.3 *Projected demand for electronics in India (Source: Boralkar, 2005)*

10.6.2 Toxic Emissions from E-wastes

Computers and other electronic equipment are made from hundreds of different materials, both naturally occurring as well as man-made (Fig. 10.4). While some substances, such as chromium, are harmless in nature, their use in the manufacture of electronic equipment often results in hexavalent Chromium, Cr(VI) which is extremely hazardous.

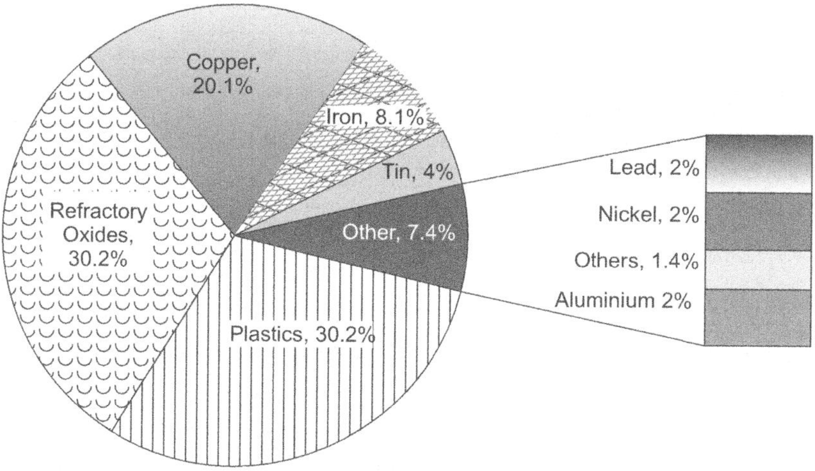

Fig. 10.4 *Composition of E-waste (Source: UNEP, 2005)*

Most nations lack specific regulations and management practices for E-wastes hence these are discarded in landfills which are susceptible to leaching of many hazardous toxicants from this waste into groundwater (Table 10.3).

Table. 10.3 *Hazardous substances in E-wastes*

Substance	Occurrence in E-waste
1. Halogenated compounds	
Polychlorinated biphenyls	Condensers, Transformers
TBBP-A (tetrabromo-bisphenol-A) PBB (polybrominated biphenyl) PBDE (polybrominateddiphenyl ethers)	Fire retardants for plastics (thermoplastic components, cable insulation). TBBA is the most widely used flame retardant in PCBs and casings.
CFC Chlorofluorocarbon	Cooling unit, Insulation foam
PVC (polyvinyl chloride)	Cable insulation
2. Heavy metals and other metals	
Arsenic	Small quantities in the form of gallium arsenide within light emitting diodes
Barium	Gutters in CRT
Cadmium	Ni-Cd-batteries, fluorescent layer (CRT screens), printer inks and toners,
Chromium VI	Data tapes, floppy-disks
Lead	CRT screens, batteries, PCBs
Lithium	Li-batteries
Mercury	Fluorescent lamps backlighting in LCDs, some alkaline batteries and mercury wetted switches
Nickel	Rechargeable Ni-Cd batteries or NiMH batteries, electron gun in CRT
Selenium	Older photocopying-machines (photo drums)
Zinc sulphide	Interior of CRT screens, mixed with metals
3. Rare earth elements and others	
Yttrium, Europium	Fluorescent layer (CRT-screen)
Toner Dust	Toner cartridges for laser printers/copiers

(Source: EMPA, 2004)

The other common practice of incineration is a poor alternative because the heavy metals cannot be destroyed in an incinerator. Most of the metals concentrate and accumulate in the bottom ash produced by the combustion process, while a portion become gaseous and exits with the flue gases. Capture of these escaping metals is a function of the efficiency of the ESP or bag house being used at the facility. Removal of mercury in any case is negligible.

At some places recycling is undertaken to extract valuable components from E-waste but unplanned and improperly regulated spawn of the E-recycling industry poses enormous occupational and environmental health threats.

10.6.3 Disposal through Landfilling and Incineration

The vast majority of E-waste ends up in landfills. According to the US Environmental Protection Agency, in 2000, more than 4.6 million tonne of E-waste ended up in landfills. It is common knowledge that all landfills leak. Even the best "state of the art" ones are not completely sealed throughout their lifetimes and a certain amount of chemical and metal leakage will occur. The situation is far graver for older or less stringently controlled dump sites. Uncontrolled fires may begin at such landfills, posing additional health and environmental risks. The vaporization of metallic mercury and dimethylene mercury is also of concern. Increasing the need for landfills is a burden to environment. Also, with a shortage of landfill capacity in populated areas and an increased concern about environmental quality, fresh waste management options are desired.

On the other hand incineration is a poor alternative because, as stated earlier, the heavy metals cannot be destroyed in an incinerator. Most of the metals concentrate and accumulate in the bottom ash produced by the combustion process, while a portion become gaseous and exits with the flue gases. Capture of these escaping metals is a function of the efficiency of the ESP or bag house being used at the facility. Removal of mercury in any case is negligible.

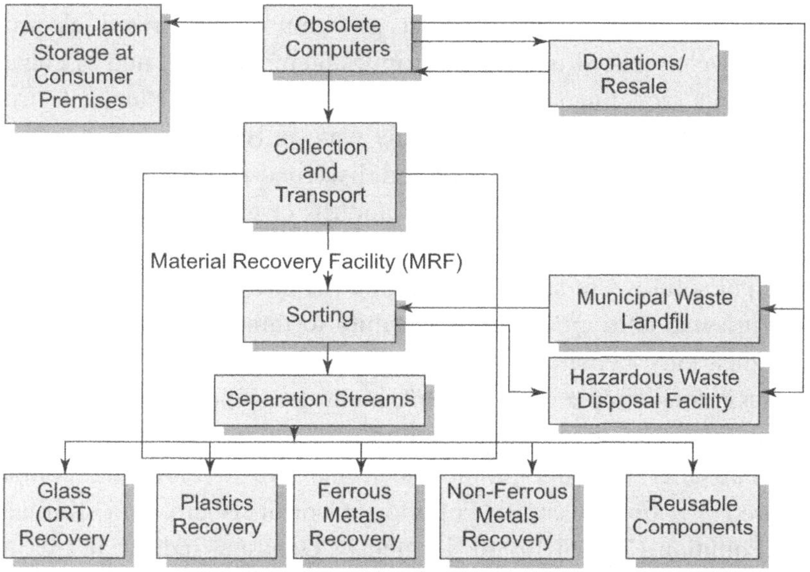

Fig. 10.5 *Material flow in recycling of E-wastes*

10.6.4 The Indian Scenario

The general practice in India is to "sell" E-wastes and such items to "Kabbadis" or "Raddiwalas" who go from house to house collecting them (for a very small monetary compensation to the householder). The material so collected is taken to a few dismantling sites where "cannibalism" starts. Materials considered valuable and reusable are picked out from the E-wastes for resale, while the rest find their way to municipal landfill sites. Often some useful items as well as hazardous ones also end up in the landfills because the whole operation is done in an unprofessional way. The material going to the landfill eventually leads to the formation of contaminated leachates which are equally hazardous and difficult to treat or separate.

Special agencies for collection and disposal of E-wastes in *a professional* manner has just come to India as described below for a Company doing this in Bangalore, recently, where all the E-waste is handled professionally by the Company and sorted out carefully. Reusable material goes to respective manufacturers while hazardous and non-reusable material goes to secure landfills. Only non-hazardous, non-reusable material goes to regular municipal solid waste disposal sites.

Company Profile

E-Parisaraa Pvt. Ltd, India's first government authorized electronic waste recycler started operations from September 2005, and is engaged in handling recycling and reusing of Waste Electrical and Electronic Equipment (WEEE) in an eco-friendly way, in Bangalore. The initiative is to reduce accumulation of used and discarded electronic and electrical equipment, which mostly end up in landfills or are partly recycled in an unhygienic condition by backyard recyclers and then partly thrown into municipal solid waste streams damaging the environment. The objective of E-Parisaraa is to create an opportunity to transfer waste into socially and industrially beneficial raw materials like valuable metals, plastics and glass using simple cost-efficient, home-grown, environment-friendly technologies suitable to Indian conditions.

E-Parisaraa is India's first Electronic Waste recycling company approved by both Central Pollution Control Board and Karnataka State Pollution Control Board, located in Dobaspet industrial area near Bangalore, in the area already earmarked for hazardous waste management by the Government of Karnataka.

Services Provided by the Company

1. Inspects and collects E-waste upon paying the generator.
2. Provides logistics support.
3. Destruction, defacing service from custom debonding and IPR (Intellectual Property Right) protection.
4. Data security: E-Parisaraa is more concerned about customer's data. (They have all the possibilities of data wiping and degaussing of hard disks, data tapes, etc., along with complete destruction certificate is provided to customers along with the pictures of the same (upon request), according to their E-waste materials and requirements respectively).
5. A destruction certificate and destruction pictures (in case of request) is provided as proof of material destruction.
6. Guides customers on legal procedures.

The process of converting a product waste into raw material includes:

1. Manual dismantling
2. Hands on segregation
3. Shredding
4. Density separation

It also includes:

1. Toner cartridge dismantling
2. Gold recovery from printed circuit Board strips and components
3. Silver recovery from silver coated components
4. Circuit for reusing fused CFLs
5. Printed Circuit Boards are shredded to 1.5" × 1.5" size and exported to Belgium for smelting

The bench mark of recycling is to recycle E-waste to about 99% with only about 1% hazardous materials which cannot be further recycled or reused and goes for scientific and secure landfill to adjacent treatment and disposal facility. Non-hazardous material goes to municipal solid waste disposal sites.

It is hoped that more such facilities will become available all over India in the near future.

REFERENCES

Asokan, P, Saxena, M, and Asolekar, S R, (2007). "Solid Waste Generation In India And Their Recycling Potential In Building Materials" *Building and Environment*, Volume 42, pp. 2311–2320.

Asokan, P, Saxena, M, and Asolekar, S R, (2010). "Recycling hazardous Jarosite waste using coal combustion residues" *Materials Characterization*, Volume 61, pp. 1342–1355.

Asokan, P, Saxena, M, Asolekar, S R, Pickering, K L, and Gupta, A K, (2011). "Design and Optimization for Hazardous Jarosite Waste Recycling in Environmentally Suitable Composite Products Using Response Surface Methodology" Chapter-2 in *Management of Hazardous Residues* (Eds. M.J. Balart Murria), 59 pages, Nova Science Publishers, Inc., USA.

Asolekar, S R, (2002). "Greening of Industries and Communities" A chapter published in the book *Rio to Johannesburg: India's Experience in Sustainable Development* (Ed. LEAD India), Orient Longman, Hyderabad, India. [Originally published in August 2002: first reprint in December 2002].

Asolekar, S R, Kalbar, P P, and Tilwankar, A K, (2011). "Framework for Sustainable Urban Environmental Services: Successful Application to High Rise Buildings in Mumbai" A chapter in the book *Designing Sustainable Products, Services and Manufacturing Systems* (Eds. Chakrabarti A., Rachuri S., Sarkar P. and Kota S.), Research Publishing Services, Singapore.

Asolekar, S R, Tilwankar, A K, and Kalbar, P P, (2009). "Towards Green Future for the Metropolitan Habitats" *Industrial Safety Chronicle, National Safety Council*, Volume XL, April–June 2009, pp. 15–17.

Kumar, S, Mukherjee, S, Chakrabarti, T, and Devotta, S, (2008). "Hazardous Waste Management System in India: An Overview" *Critical Reviews in Environmental Science and Technology*, Volume 38, pp. 43–71.

Lagrega, M D, Buckingham, P L, and Evans, J C, (1994). *Hazardous Waste Management* McGraw-Hill, New York, USA.

Code of Federal Regulations (40 CFR) 260 (dealing with *Collection, Packaging & Transportation, Storage, Treatment and Disposal of Hazardous Wastes*) published by the US Environmental Protection Agency, USA during 1987–2012.

Recommended Additional Reading

Regulations passed by the Ministry of Environment and Forests (MoEF), Government of India, New Delhi as listed below:

- Manufacture, Storage, and Import of Hazardous Chemicals Rules, November 1989.
- Hazardous Waste (Management and Handling) Rules, December 1989.
- Rules on Hazardous Micro-organisms/Genetically Engineered Organisms or Cells, December 1989.
- Hazardous Wastes (Management, Handling and Transboundary Movement) Rules, 2008.
- Municipal Solid Wastes (Management and Handling) Rules, 2000.
- Bio-Medical Waste (Management and Handling) Amendment Rules, 2003.
- Batteries (Management and Handling) Amendment Rules, 2010.
- E-waste (Management and Handling) Rules, 2011.

EXERCISE QUESTIONS

1. What are hazardous wastes? How are they defined in the "Indian Regulations"?
2. What are the characteristics of a given waste which help in deciding whether the waste is "hazardous" or not?
3. What is the Toxicity Characteristics Leaching Procedure (TCLP) approved by the US Environmental Protection Agency (US EPA)? When is the test typically conducted? What does it address and what are its limitations (or drawbacks)?
4. What are the reuse options for different wastes particularly those having low toxicity and high volumes? Name the waste stream and describe the application categorically.
5. When would one practice solidification and stabilization of the wastes? Name the waste stream and describe the application categorically.
6. You are aware that one of the ways of ascertaining if the waste matrix is hazardous is to analyze it and quantify certain properties such as: flammability, explosivity, toxicity, corrosivity, reactivity, etc. If you subject the commonly encountered "road tar", you will

find that it contains dozens of carcinogenic toxic compounds in large quantities, why should we not then consider road tar as hazardous?

7. What are the major municipal solid wastes (MSW) on our academic campus and hostels? What management solutions and what technologies are there to collect, transport, treat and dispose them off?
8. There are four major E-wastes on our campus – produced by students and academic departments, namely:

A. Fluorescent Tube Lights (FTLs) and Compact Fluorescent Lamps (CFLs)
B. Old CRT monitors from discarded PCs
C. Laptops and Cell phones
D. Dry cell batteries, phone batteries and laptop batteries

Give your commentaries on each one of the above wastes in the following tabular format:

[i] Major pollutants,	
[ii] Impact on health,	
[iii] Impact on environment; and	
[iv] Technology and/or Management strategy for safe disposal.	

PROJECT/ASSIGNMENT TOPICS

Hard Options

Instructor can assign group/Individual projects on the following topics to study industrial sectors (or an environmental issue) as a focal point for listing hazardous wastes and describe their characteristics:

1. Production of sponge Iron
2. Natural gas production
3. Production of Zinc
4. Production of Copper
5. Production of Lead
6. Production of aluminium (primary or secondary production)
7. Production of chlorine by means of mercury/diaphragm-electrolyses process
8. Phenol production

9. Production of fertilizer
10. Production of solvents
    (Cracking of Naptha and cyclisation unit of integrated petroleum refinery)
11. Production of coatings paints, lacquers, varnishes and plastics, and of inks
12. Production of cements
13. Production of glues, adhesive and resins
14. Production of dyes, dye-intermediates and pigments
15. Production of halogen-containing hydrocarbons or of aromatic, aliphatic or napthenic hydrocarbons
16. Production of pharmaceuticals
17. Production of pesticides
18. Tanning of leather
19. Production of Antibiotics

Soft Options

Instructor can assign group/Individual projects on the following topics to study following hazardous waste treatment / management technologies:

1. Air/steam Stripping
2. Wet Air Oxidation
3. Supercritical Fluid Extraction
4. Carbon Adsorption
5. Membrane Processes
6. Solidification / Stabilization
7. Incineration
8. Plasma Incineration
9. Pyrolysis
10. Ex-situ Soil Washing
11. Soil Venting
12. Solvent Flushing
13. Engineered Landfilling

11 CHAPTER

Some Useful Initiatives

11.1 Some Corporate and Community Initiatives

11.2 Low Carbon and "Greening" Efforts at Personal and Community Level

11.3 "Green" Buildings

11.4 "Green" Transport

11.5 "Green" Infrastructure for Municipal Services

11.6 "Green" Industries

11.7 Ensuring Sustainability in Whatever We Do

11.8 Case Studies

11.9 Conclusion

We always hope our initiatives will someday hatch into innovations and eventually become common practice in course of time.

11.1 SOME CORPORATE AND COMMUNITY INITIATIVES

Let us consider corporate initiatives first. Those who run a company have to take many initiatives from time to time to keep it running well. Besides operating the Company efficiently and profitably as far as its main production is concerned, a few other Corporate activities have to be undertaken from time to time in support of the main activity. Most of these activities are undertaken either to meet government's mandatory requirements, or industry's expectations, or to improve the Company's bottom-line and shareholder value.

Some of these activities are listed below and speak for themselves. The reader is advised to refer to the internet for further details on each of these items:

- Adoption of good house-keeping practices
- Adoption of cleaner processes and operations
- Audits for energy consumed, water consumed and raw materials used
- Audits for wastes produced (solid, liquid and gaseous)
- Noise abatement, heat and stress abatement
- Control of flies, mosquitoes and rodents
- Monitoring quality of air, water, fuels used
- Cost of treatment
- Staff health, welfare and safety
- Greening of the industry (see next section)

11.1.1 Adoption of Good House-keeping and Cleaner Processes

There are many interesting examples of Cleaner Processes which can be adopted to reduce costs and consumption of valuable resources. Industries wanting to adopt Cleaner Processes have to adopt at the same time good housekeeping which will conserve water, chemicals and fuels used in manufacture, segregate wastes and undertake in-plant modifications to facilitate waste treatment, reuse and recycle, by-product recovery, etc.

All such activities cost money and time. But, this often constitutes money well spent as benefits flow from it. The first net benefit of adopting above type of measures is that overall costs decrease, which directly improve the bottom line of the company. A second benefit which often results from such action is that staff health and welfare also improve which increases general productivity and thus again improves the bottom line. One more factor helps the bottom line to look better, i.e., income from by-product recovery.

Audits and Waste Reduction

Energy audit, water audit and waste reduction studies (solid and liquid) are all relatively inescapable activities which have to be undertaken in recent times to meet the demand for holistic approaches to be followed in production planning and operational activities.

Industrial groups are beginning to get aware of the need to retrofit themselves to reduce especially electric power and water consumption in their factories, because of their beneficial effect on their company's bottom lines and on global warming as well as resources conservation. Newer technologies are sought whenever new installations are being planned.

Monitoring of Costs and Quality of Raw Products Used

This would be done in any commercial operation to control expenses and quality of various raw products used. Careful records have to be kept which would enable conclusions to be drawn and strategies refined.

Corporate Social Responsibility (CSR)

This section on Corporate Social Responsibility has been included knowing that all engineers and others who join industries or public sector undertakings are expected to know something about CSR activities as they affect the company's working.

The term "corporate social responsibility" is pretty well known in the industry. It is an important part of Corporate Governance. It is at present a mandatory requirement for all public sector undertakings and (as yet) only a recommendatory one for private industrial bodies to set aside a small fraction of their annual net income (upto 2%) for undertaking such activities. The idea seems to be to induce the corporate sector to show a greater sympathy towards the weaker sections of society. It should not be that only the government should take care of the weaker sections and the industry only increases shareholder value. All have to do their bit.

A beginning has now been made on a country-wide scale by Government with public sector undertakings and willing industries from the private sector.

> Shouldn't industry also show greater interest in the welfare of the country's weaker sections? After all, India has the largest number of poor people in the world, and Government alone cannot do everything.

Tata Iron and Steel Company are reputed to have made a beginning with Corporate Social Responsibility from early 1900's in their steel town, Jamshedpur, where the town's infrastructure and various amenities have been set-up keeping in mind health and welfare of their staff and families. Their coal and iron ore mines located in Naxalite infested areas are even today allowed to function without hindrance as the local people have experienced the benefits of their CSR policies.

Already some private sector industries have started to undertake CSR activities with pride. The good sign is that their number is gradually increasing as they find that their CSR policies are winning the confidence of the local people, and of their own shareholders. One of the important benefits to accrue from this effort is the increasing attention being given to "dalit" causes as a result of adopting policies which benefit all their staff (inclusive policies).

A way has, however, to be found to include the large number of small and medium-scale industries in the unorganized sectors (who are too small to have an effective CSR of their own to uplift the poor). A joint effort much like the effort which has led to the installation of Common Effluent Treatment Plants discussed in Section 7.6 is perhaps required.

The core element of CSR activities usually includes the following:

- Ethical functioning of the company (reflecting the company's ideals) in keeping with public's expectations.
- Respect for health and welfare of all stakeholders affected by industry (not just increased profits for shareholders).
- Care for the environment.
- An inclusive growth and education for all (including scheduled castes).

Of course, each industry's sense of values can be different, but the three main streams of activities often embraced in the social field are: *education, environment and health care*, including safe drinking water and sanitation. To enable people to focus on these items, income generation, education and sustainable livelihood programs have to be first encouraged. These programs have to ensure inclusive growth, and not be confined to a select few, in order to win all round sympathy and participation.

The programs mostly start by undertaking field surveys to establish the present status followed by educational programs to promote certain aspects of their work. Alternatively, companies undertake health surveys followed by health or environment improvement programs. In practice, companies like to undertake projects which have a relatively small gestation period and benefit large number of people (generally their own workers and their families). The results are reported with pride at the end of the year in their annual reports. Most parties prefer to take up either education or health oriented projects which in one way or another affect the environment – and are good for their own employees and their families.

In recent times, some companies have also devoted attention under CSR to process changes and operational changes for reducing emissions of CO_2 and other greenhouse gases from their factories and thus try to secure UN's Carbon Credits for such work under the Kyoto Protocol.

Corporate "Irresponsibility"

A word of caution needs to be mentioned here: Just as there is Corporate Social Responsibility, there can also be Corporate Social Irresponsibility! One can never balance the two. Irresponsibility can be shown in many a way, and lead to a trust deficit of wide magnitude which can even affect the product's brand value. Examples are:

- Use of unethical or illegal methods in manufacture.
- Adulteration of product.
- Unethical, wrongful and mischievous advertising.

Adulteration and wrongful advertising are to be particularly guarded against in the food, nutrition and pharmaceutical businesses, some of which have also harnessed the sciences and statistics to promote their product or to wrongfully misguide the public through clever use of defective or questionable research cloaked in even more questionable statistics. The questions one could ask are:

- Are we comparing apples with apples or apples with oranges?
- Is the study published in a reputed, peer-reviewed journal?
- Is the finding supported by any independent evidence?

- What was the duration of study and size of sample?
- Who funded the study?, etc.

Such questions can be asked to discover the true intent of a study or piece of so-called research. The purpose is not to doubt the validity of each and every piece of work but to guard against clever fraud. Generally, most reputed and professional journals which are peer-reviewed can be trusted.

> Industries are aware that it takes years and years to build up a respectable "brand value", but the same value can be lost within a very short period by some wrong activities.

Community Initiatives

Among community initiatives shown in recent times for improving the environment, mention must be made of the following:

- "ALMs" efforts started in some cities. ALMs is an acronym which stands for "Area Local Management". People have often got together (Public-Private-Participation) to improve the looks or cleanliness of an area which languished under local municipal management – a kind of community awakening to form a local group to look after an area's sweepings and other activities allegedly neglected by the local authority. Whole areas have got transformed and beautified under local leadership – the idea being not to replace the local authority but to supplement its efforts to fulfill its tasks.
- Launching a Public Interest Litigation (PIL) and achieving success, is another way of demonstrating a community initiative in the Indian context.
- The internet and web services which are freely available to those having access to a computer. These services have a wealth of knowledge which can spark useful ideas.

A community initiative has to be initiated by community action. When a local body (or municipality) takes some action of its own accord, it does not fall in the category of community initiative. For example, a local body may decide to augment a water supply, or undertake water conservation practices or treat wastewater before discharge or treat and reuse it for other purposes, etc. This is not considered to be a community initiative, but an official initiative of the local body.

11.2 LOW CARBON AND "GREENING" EFFORTS AT PERSONAL AND COMMUNITY LEVEL

As already mentioned in Chapter 2, emission of greenhouse gases including their major component carbon has to be reduced. The world needs to cut down on its use of fossil fuels like coal and oil because their use involves an emission problem for all countries (and a resource problem for many countries, such as India). We have to think "*green*" at every level.

Every country has to undertake programs for reducing their "carbon footprints" by reducing carbon "sources" and increasing carbon "sinks". Such programs include:

- More efficient use of electricity and other fuels in our homes, in transport and in industry in order to reduce emission of carbon upon combustion.
- Increasing use of alternative sources of energy which do not emit any carbon upon combustion.
- Protection of existing "sinks" (such as forests) and promotion of new one to absorb carbon from atmosphere.

Hence, in this chapter we start with lifestyle changes and go on to green buildings and area planning. Then, we go on to think of green transport, green industries and a green municipal infrastructure for city services because thinking "green" helps to:

- conserve our resources (fossil fuels, water, forests, etc.).
- reduce our expenditure on use of costly electricity and oil.
- generate less pollution (including greenhouse gas emissions) from various activities.

A few examples are given below (not in any order of importance) which would conserve the use of electricity and may in some cases even postpone need for additional power generation capacity in an area. What is more directly beneficial is that lesser electric bills would have to be paid by the people concerned, and lesser emission of CO_2 would take place to the atmosphere:

In Housing

1. All public hotels, hostels, hospitals, and large residences must be required to provide solar water heaters for their bathrooms and kitchens.

2. Use less A/C which is the largest single consumer of electricity. For existing buildings, at least set A/Cs (thermostats) to kick in at slightly higher temperatures (say 24–26°C) to save electricity.
3. For all new equipment check its energy efficiency rating as given by Bureau of Energy Efficiency (BEE) and buy the most efficient equipment and appliances when buying new equipment.
4. Change over to CFL type of bulbs in all offices, homes and shops as they require much less electric power than incandescent bulbs for the same illumination.
5. Use timer switches to switch-off lights when not needed in common areas like staircases, passages, etc. Timer switches have just not been used in India so far. Overseas, they are commonly used.

Transport

1. Go by bus or train instead of plane for short journeys. To avoid longer air journeys, substitute video conferencing instead of air travel, if possible.
2. Send children to school by bus instead of private car. If car must be used, carry a few other passengers too. Car-pooling is a good idea.
3. Use a smaller car as it has lesser emissions. One might have to plant a fewer number of trees to be able to offset the CO_2 produced by one car. Use CNG (compressed natural gas) instead of petrol. Thousands of taxis in Mumbai and some cities of India already operate on CNG, which is a relatively cleaner fuel than diesel oil or petrol.
4. Washing a car with a hose consumes about 400 L of water whereas washing it with a bucket only needs a bucketful or two of water.

Miscellaneous

1. For street lights, traffic lights, flood lights and for large, lighted-up outdoor hoardings (advertisements) use Liquid Crystal Display (LCD) or better still Light Emitting Diode (LED) type fixtures to economise on power consumption. (Punjab stands out as the first state in India to introduce LED lighting in 5 cities).
2. Use less plastics or use plastics made from renewables such as Poly Lactic Acid (PLA).

3. Use locally available foods, construction and industrial materials as far as possible, so as to cut down on CO_2 emissions in their transport from the country of origin.

Sweden has already cut its Greenhouse Gas emissions by 17% since 1990. The city has reduced its use of fossil fuels from 70% in 1970 to 30% today, and has a plan to be completely fossil-fuel free by 2050. By the year 2020, its entire fleet of buses will be operated on renewables. Stockholm's heating is already based on bio-fuels and the entire food waste is proposed to give biogas.

11.3 "GREEN" BUILDINGS

Green buildings can save a community a lot of fossil fuels and carbon emissions. Green buildings also help conserve other resources such as water. They use local building materials as far as possible, and recycle waste and wastewater.

> May be, a "green" building might cost 5–10% more in capital cost, but it is certainly less expensive in operating cost. Thus, overall it becomes a more economic proposition: the additional capital expenditure is made good in a few years through savings in operating expenses.

Green buildings reduce a building's carbon footprint forever afterwards and reduce its demand on a country's natural resources. The scope for reducing carbon emissions is high because buildings are said to account for 40% of all energy used in this world today and energy demand is expected to increase by 30% within the next 10 years. Urbanization is so rapid that 60% of world population is expected to be living in cities in the next 15 years. So green buildings and green infrastructure are must.

11.3.1 LEEDS and GRIHA Rating Systems for Green Buildings

To have what we call a "green" building, one has to pay attention to the following guidelines set by the "Leadership in Energy and Environmental Development Systems" (LEEDS guidelines). TERI, New Delhi, and the Government of India have together developed a similar Indian version called Green Rating for Integrated Habitat Assessment (GRIHA) which is said to be more suited to Indian conditions and approved by Government of India. Other organizations in the world have also prepared such guidelines for their use.

Essentially, the LEED or any other guidelines, helps to develop a stronger sensitivity towards use of energy-consuming devices and a better understanding of the environmental impact of the developers' actions. Climate and nature must be respected in everything we do.

For example, it is a well known fact that owing to latent heat far greater energy is required in cooling or air-conditioning a space than in heating it. Hence, in a country like India, one has to be extra-careful to avoid aggravating the heat by wrong architectural planning and then using more electricity to overcome the situation by air-conditioning. Excessive use of air-conditioning is not good economics and must be avoided by better planning and choice of building materials, etc.

Some guidelines for design of buildings in India have been available since time immemorial when building designers were taught that in India there are two major types of climate situations: (1) Hot-humid, as in South India; and hot-dry, as in North India. Building design has to be distinctly different in the two cases. Modern architects sometimes blur these natural distinctions by using similar architectural design everywhere and making up by providing air-conditioning. Similarly, designers sometimes provided "modern" features without ascertaining first whether the required infrastructure exists. Waterless urinals, for example, cannot be provided where power failures often happen.

11.3.2 LEEDS Guidelines

LEEDS guidelines cover various aspects of a project design starting with the building construction site itself and proceeding up to the various planning modules and aspects as listed below including social relevance without which it would be an empty effort.

All new "green" buildings are required to comply with the guidelines. Some larger, old buildings under major renovation could also be included, at least to some extent. The guidelines are really based on commonsense matters and, if followed, would help to make life better, cheaper and in line with modern environmental concepts. The following text summarizes in brief what can be done under each item:

1. The Building Construction Site

- ➢ Design with minimum disruption on site; minimize soil displacement; minimize soil erosion. The more the soil is displaced, the greater is the likelihood of upsetting the natural drainage.
- ➢ Preserve and reuse nutrient-rich top soil for landscaping.

- Reduce micro-climate temperature rise by planting shady trees.
- Minimize pumping; let drainage follow existing slopes/contours; facilitate easy maintenance.
- Preserve biodiversity, compensate by reforestation if necessary.
- Facilitate groundwater recharge; restrict run-off and local flood problems.

2. Environmental Concerns in Architectural Planning

- Achieve thermal comfort (e.g., use hollow blocks for walls to keep houses cool and cut down on use of A/Cs). Use of hollow blocks, double-wall construction and other methods should be preferred from insulation viewpoint in India, rather than use of ordinary bricks.
- Avoid use of glass facades especially on sun-side as they greatly increase the load on air-conditioning/electricity. (Glass is OK on north side in Mumbai). Many buildings unfortunately use architecturally attractive full length glass panels on the side facing the road (even if it happens to be on the sunny-side). This places a higher load on the air-conditioning system of the building, and hence on expensive power consumption.
- Achieve visual comfort (through choice of colors, materials, etc.).
- Prevent heat gain through use of larger roof overhangs to extend shade for longer hours. Scientifically designed vertical and horizontal fins (sun-breakers) could be provided in some cases. Double-glazed window panes would also reduce heat transfer. Also, use double-roofing for heat insulation where a sloping concrete roof is covered over with Mangalore or Spanish tiles with an air gap inbetween to protect the concrete roofing from the fierce sun.

(Since hundreds of years, use of concrete or stone "jalli" and a lightly-flowing water fountain has been an effective device used in the hot and dry climate of North India (Example: Taj Mahal, Agra).

However, ensure entry of adequate daylight inside the house to avoid use of electric lights during day time (important to use skylights, transparent sheets, etc., which would admit natural daylight and save electric power).

3. Energy Conservation and Better Management

Heating, ventilation and air-conditioning (HVAC) are usually the three heaviest users of electricity. Thus, some attention needs to be paid to the efficient use of electrical energy by them. Of course, their original technology also determines their efficiency. Normally, air-cooled window air-conditioners are said to use more electricity per tonne of refrigeration capacity than equally large units with re-circulating waters. For air-cooled units the electrical consumption is said to vary from 1.6 to 1.8 kW per tonne of refrigeration capacity. The coefficient of performance for water-cooled chillers varies from 0.72 to 1.0 kW per tonne. Their maintenance cost is further reduced by providing double-glazed window panes, sun-breakers, etc., and ensuring thermostat control of temperature, as recommended above.

LEEDS Power Recommendations

LEEDS recommends to restrict light power density to 7.5 W/sq m or less, as far as possible in a green building. ECBC also makes useful recommendations suitable for India (see below).

The "TERI" Method

TERI, New Delhi has demonstrated for Delhi's hot-dry climate: how use of electric power for air-conditioning can be saved by providing air tunnels of a certain dimension at 4 meter depth through which air is blown.

At that depth in Delhi, the air is literally uniformly cool throughout the year, and so the air blown through it either picks up heat or discharges heat depending on the season. As everybody knows, a blower consumes much less power than the compressor of an air-conditioner. Moreover, the reduced power is met from a renewable energy source.

Hence, the building consumes about 1/3rd power of what a conventional building would consume) in Delhi.

4. Water Conservation

- Minimize use of public water supplies. Conserve water. Reuse wherever possible.
- Provide for rainwater harvesting and groundwater recharge to minimize use of water from public water supplies.

- Provide low-flow fixtures, orifices in pipes and use dual-flushing tanks to minimize use of water. Provide waterless urinals where power supply is dependable.
- Reuse grey/black waters after suitable treatment for flushing or gardening.
- In gardens, use plant species which inherently require less water.

5. Waste Management

Use a "natural" method of treatment such as a lagoon, pond or land irrigation which does not need use of electric power which would otherwise be needed in case of a mechanized method for aeration, pumping, etc.

Reuse wastewater as far as possible in gardening/crop irrigation/ groundwater recharge/other uses at site, after appropriate treatment.

Reuse solid wastes after segregation to salvage re-usable materials and be able to compost organic wastes and recover biogas and manure for use in garden.

6. Ensure Social Relevance

- Ultimately, whatever is done must have social relevance and be good for the local people. We must use affordable, durable and low maintenance building materials which are locally available, need locally available skills and management systems to maintain and avoid use of fuel for transport of men and materials from distant countries.
- Reduce consumption of all resources. Avoid wastage of materials and production of all forms of wastes.
- Promote reuse of water.
- Use green plants to keep up air quality (e.g., Areca Palm, a living room plant helps oxygenate during day time while sansvieria trifasciata helps oxygenate during night time).
- Adopt preferential financing of "green" projects by home loan financing bodies.
- Develop supportive micro-enterprises. Promote environmental sustainability.

(Mumbai Municipal Corporation is introducing a welcome inducement by reducing the Property Tax on green buildings).

Green Building Ratings from Energy Consumption Viewpoint

Most commercial buildings in India have an Electrical Performance Index (EPI) of 200–400 kWh/sq m /year whereas similar buildings in Europe and USA have an EPI of 150–200. Good design in India can reduce EPI to 100–150 but Indian builders fear that they would bear increased capital costs while the tenants would benefit from the lower operating costs!

Most of the electrical consumption in a building (over 55%) occurs in heating, ventilation and air-conditioning (HVAC). The remaining is consumed in lighting (14%), electronics (24%) and miscellaneous uses (4%).

> While the benchmark is 180 kWh/m^2/year, ECBC compliant buildings average 110 kWh/m^2/year. The benefit of reduced electrical consumption is now being obtained by many buildings in India.

The first building in India to be certified as "Platinum" under LEEDS system has been the CII-Sohrab Godrej Green Business Centre at Hyderabad. Several other buildings have earned similar ratings from LEEDS and GRIHA. These ratings are now being extended to cover different types of buildings (hotels, hospitals, IT Parks, malls, etc.) and in different climatic conditions.

Second Generation Green Buildings

The green buildings we saw above may rightly be called first generation green buildings. They used various architectural and engineering devices to reduce the electricity, water and other resources needed to operate the building.

Second generation buildings, according to the authors, not only reduce requirements as above but also generate their own electricity (in part or full) from a renewable source like wind or solar energy or both (hybrid). This electricity is either used in the building itself or fed-back to the local electric supply grid company at specially subsidized rates fixed by Government to make the transaction attractive. The local grids in India are required to buy back up to 5% of their total electric supply. (Germany requires up to 15 %).

Builders selling ownership flats can provide wind-mills and solar panels on the building terrace to generate some electricity which can be supplied free to the building owners to supplement the electricity they buy from the city grid. In this way two things would happen: Firstly, the builder would find it easier to sell a flat where some electricity is provided free, and secondly, a kind of public-private partnership would occur for funding electricity generation.

> **Observation:** A scheme can be developed on above lines for a middle-class building in a selected locality to demonstrate its workability. A visit could be organized to a green building if necessary to obtain the required data.

11.4 "GREEN" TRANSPORT

The many varieties of transport used in a city or country can all be placed under two broad groups:

- Private modes of transport such as cars, taxis, SUVs, etc., and
- Mass' modes of transport such as trains, buses, underground metros, monorails, coastal ships, etc.

It is clear that the "greening" of any city's transport system is invariably affected by fuel pricing policy and the type of transportation available to the people. Whatever the fuel used, emission data shows that per passenger-kilometer travelled, a private mode of transport always gives a higher carbon emission than a mass transport system. In spite of higher emission, private mode of transport is often preferred for various reasons.

Yet, it must be remembered that private transport also has an inherent limitation as it occupies road space and traffic speeds tend to become slower and slower as time goes in spite of flyovers and road over-bridges. This depends on the town-planning that exists, and cannot be undone. In a country like India where population growth is rapid and urbanization is even more rapid, broad roads, flyovers, elevated highways and such provisions are useful but only up to a point.

Another problem with crowded Indian cities is that when any area is newly developing with an industry it causes an "induced" growth of people employed in the industries creating a new slum as they want to stay close to their site of work. Additionally, new industries generate more passenger and goods traffic in their vicinity.

Private transport, in its effort to survive under urban conditions, is developing the hybrid car and better still the electric car with "zero" emissions. The latter will need a better car battery and the car battery will need to be recharged every night by using renewable energy produced in the building itself from wind and solar systems. This is all technically possible and will be welcome as no petrol, no diesel, no bio-fuel would be required! Car maintenance would also become easier. This is a very feasible scenario and one can be sure it will happen some years down the line although it will be resisted by vested interests.

Mass transport facilities will, however, be needed in the future (not because of their lower per capita emission) but mainly because there is a limit to the build-up of vehicular traffic that can be handled by any city's roadway system, and every city would like to attract more and more people to come and live in it for commercial and business reasons. Hence, both forms of transport will always be needed and have a great future in every city.

Blessed are the old world's cities that installed mass transport systems (underground railways, tubes, metros, etc.) in the good old days when costs were low and cities were not so built up. Today the costs are much higher, the cities are more built up and traffic is heavy. Yet, we have to include mass transport in our planning because individual transport and city roads have their limits, and newer technologies make mass transport more feasible, and in some cases the only way out. In coastal cities where water transport is feasible, it should also be considered.

> There is an American saying that: "If all you ever do is all you have ever done, then all you will ever get is all you ever got". We have to re-invent ourselves from time to time.

11.5 "GREEN" INFRASTRUCTURE FOR MUNICIPAL SERVICES

Green infrastructure covers various municipal services and refers to almost the entire infrastructure consisting of its

- water supply distribution network,
- storm water collection and disposal network,
- wastewater collection, treatment and disposal system,
- solid wastes collection and disposal system,
- industrial and other hazardous wastes disposal system.

In each item, much emission of greenhouse gases occurs and therefore, there is good scope for some "greening" to be done as briefly indicated below. This subject is now receiving increasing attention, and newer and better ways of control will surely emerge in the future.

11.5.1 Water Supply and Distribution

From the viewpoint of reducing CO_2 emissions, electricity usage must be minimized in water treatment and water distribution, since the production of 1 kWh of electricity releases 0.44 kg CO_2 at the power generating station.

Indian farmers are reported to use 75–85 billion kWh of electricity and 3.5–4.0 billion litres of diesel oil per year in pumping groundwater. These are frightening statistics. Deep tubewells in Gujarat, Rajasthan, Punjab and western UP are great power guzzlers. Groundwater irrigation accounts for 4–6% of the country's carbon emissions. A lowering of 1 m in groundwater level increases carbon emissions by 4–6% from pumping. An increase of 1% in groundwater irrigated area increases emissions by 2.2%.

A gravity flow scheme must invariably be preferred rather than a pumping scheme, if a choice exists. Where pumping is involved, pressures in the system must be optimized so as to meet desired pressure criteria without necessitating unduly heavy pumping.

> Unaccounted-for-water (UFW) losses from distribution systems in Indian cities are often unacceptably high (>20%) and need more serious repair-and-maintenance routines by local body staff. Leakages obviously mean a corresponding volume of water has to be pumped additionally.

11.5.2 Storm Water Run-off

With big changes occurring in rainfall pattern and sea levels rising as a result of global warming, the functioning of storm water drains is going to be severely impacted in all coastal cities. Wherever the natural ground slope between midtown and the coastal waterline is not enough for drainage to take place by gravity, pumping will be required.

Mumbai is already setting up its first storm water pumping station at Haji Ali as level difference is not enough especially at high tide. This will place a demand on the power supply grid when the pumps are operated since a separate renewable source of energy has not been provided.

All storm water drainage systems of coastal towns will need to be reviewed in the light of local rainfall and runoff data every few years and suitable measures taken to provide, for example, flap gates on the outfalls to prevent in-flow of sea water into storm water drainage outfalls at high tides, dykes to create holding ponds, provision of large volume pumping stations, large canals, locks, etc., almost all of which we did not need all these years. Someday, one may even need to learn from Venice and Holland in this regard. All of these solutions will involve use of electricity and hence lead to increased emission of GHG unless a renewable source of energy is used.

11.5.3 Wastewater Collection, Treatment and Disposal

For the purpose of illustration, without going into details, we note that aerobic processes like Activated sludge need a lot of power. Hence, they should be used only where other processes such as natural systems will not meet the requirement of final BOD. In our warm and sunny climate we can also have other options such as the UASB, and the oxidation pond, both of which when used judiciously can give good results and save us power as well as other operational difficulties.

Site selection of treatment plant location should as far as possible be such as to avoid pumping of the wastewater either before or after wastewater treatment.

> An activated sludge plant to serve 1 lakh people in a typical mid-sized Indian town, requires 15 lakh kWh of electric power per year which releases an enormous quantity of 1800 kg CO_2/day or 657 tonne of CO_2 per year at the power generating station.

11.5.4 Solid Waste Disposal – Municipal, Industrial and Hazardous

Maharashtra is reported to have a coverage of only 20% as far as municipal solid waste collection and disposal is concerned. Since much work has yet to be done, solid waste collection and disposal need not be done in the conventional way, but rather in a more effective and economical way, which will also avoid global warming. For example, the trucks used for collecting solid wastes from house to house could be operated on CNG or on electricity generated from biogas, instead of being operated on diesel oil.

Municipal solid wastes disposal should also preferably be done aerobically (e.g., by aerobic composting) to stabilize it without producing CO_2. If the waste has to be treated anaerobically, the gases produced (CO_2 and CH_4) should be collected and used as fuel as discussed at length in Chapter 9.

Similarly, industrial and hazardous wastes have been discussed in Chapter 10.

11.6 "GREEN" INDUSTRIES

Industries often constitute a big part of our so-called "carbon footprint" from that area. Many industries generate carbon emissions right from the start, from ore-mining to manufacturing and continue to emit during transport and final use of the products and even during their disposal after useful life is over. Industries therefore can produce emissions "from the cradle to the grave of a product!".

From the Green House Gas (GHG) emission point of view, three industries are considered as important and widespread industries in India as their products are widely used in construction and manufacture everywhere.

- ➢ Cement
- ➢ Steel, and
- ➢ Brick-making

Other industries with significant emissions are the various food and fertilizer industries, aluminium manufacture, pulp and paper.

> The Confederation of Indian Industries (CII) estimates that there is yet some energy saving potential in Indian industry ranging from 10% in the iron and steel and aluminium industry to 15% in cement and 25% in pulp and paper.

There is likely to be a similar energy saving possibility in almost all other industries too. Moreover, besides energy saving possibility, there is always some scope for saving on carbon and other greenhouse gases emissions (such as methane) from manufacturing. It is generally wise for the industry to engage expert consultants to study the specific local situation regarding energy consumption and actual manufacturing processes used and recommend cleaner technologies and best available practices for reducing emissions.

A good example is the emission control being practiced at Tata Iron and Steel Plant at Jamshedpur where the CO_2 emissions have already been brought down to 1.8 tonne/tonne of liquid steel produced when the industry benchmark is 1.5. Many Indian industries have benefited from the UN Kyoto Protocol by finding ingenious ways of reducing their carbon emissions and selling these "certified emission reductions" to western countries at market rates.

Some Encouraging Developments into the Future

All over the world countries (including India) are taking serious note of carbon emissions and climate change and are doing things in particular such as:

- Using electricity more efficiently so as to cut down on carbon emissions as explained above.
- Setting up large-scale or full-scale city-wide demonstration projects to reduce emissions.
- Promoting research in alternative sources of energy.
- Offering incentives to industries and to the general public to reduce their electric consumption and carbon emissions.

Among city-wide demonstration projects, the efforts of several countries must be mentioned: Thailand is setting up a "Bio-solar Home" designed to be self-sufficient in all respects and not needing municipal services. China is setting up a model low-carbon city in Turpan in Xinjiang Province. Japan will do likewise in Fukuoka. Abu Dhabi is setting up a Green town of 50,000 people in Masdar with zero emissions. And New York is planning a new NY to curb their emissions.

India too is attempting to plan for reduced emissions from it new townships, for example, one at Amby Valley at Lonavla and another at the new township of Lavasa near Pune, both in Maharashtra. In both cases, the planners aim at reducing carbon emissions from wastewater treatment and solid wastes management, as well as from local traffic, housing and other sources. These examples make one feel surer that India will not lag behind in such efforts.

11.7 ENSURING SUSTAINABILITY IN WHATEVER WE DO

Sustainability is what everybody wants in whatever they do. At the outset, let us make it clear that what follows is applicable to wastewater treatment. It may be applicable to other fields also, but our primary purpose is to apply it to the environment and infrastructure in general. As common sense goes, sustainability in anything is achieved when resources continue to remain available and people continue to look after the operation. Common sense also tells us that normally people do things if they get something out of it. This is generally true in most cases.

Sustainability has two major dimensions to it – resources and people. Ensuring sustainability therefore requires two sets of steps to be taken: The first step involves adopting cleaner technologies so as to get our resources to last longer and affect the environment as little as possible and the second set consists of steps to involve the people through what is called social engineering. Both these steps are needed to be taken to ensure sustainability.

The objective of using "cleaner technologies" is mainly to reduce the use of natural resources (such as water, fossil fuels, coal and oil, minerals, metals, etc.) because natural resources are finite and have their own slow rate of replenishment. If used too rapidly they get exhausted. If used carefully, they remain available over a long period of time. Cleaner technologies also include steps to reduce waste production and reduce any ill-effects they may have on the environment.

The objective of "social engineering" is to educate the people to the benefits of the process, to create values for them, to evaluate benefits received and to ultimately pass on the benefits to the people involved so that they continue to do whatever is necessary to keep the process going. It is logical that for any scheme to be sustainable, resources must continue to remain available and people must make effort to keep the system running. The people most likely to do this are generally the beneficiaries themselves.

Hence, we require a "model" for our actions such as the one given below which in short emphasizes (1) conservation of resources, (2) adoption of cleaner technologies, (3) mitigation of unfavourable impacts, followed by (4) promotion of awareness and education, (5) increase in revenue through resource recovery, and (6) internalizing the monetary benefits among the workers to ensure sustainability.

The Proposed Model

On an overall basis, to improve sustainability, our model proposes two major approaches: "Cleaner Technologies" and "Social Engineering" with each approach having 3 sub-approaches as shown below:

Cleaner technologies

- ➢ Conserve resources
- ➢ Reduce, recycle, reuse
- ➢ Minimize all impacts

Social engineering

- ➢ Promote awareness and education in people, and in staff
- ➢ Increase revenue through resource recovery, etc.
- ➢ Internalize monetary benefits among staff

It is to be noted that both Clean Technologies and Social Engineering are equally important to have, and both must be included together for ensuring sustainability. It is not an "either-or" situation.

> For any project, the first approach is to minimize all inputs and impacts. The second approach is to maximize all benefits to get self-motivation on the part of the staff. The staff must be self-motivated to keep things running. The sum total of this effort is to increase sustainability of the project.

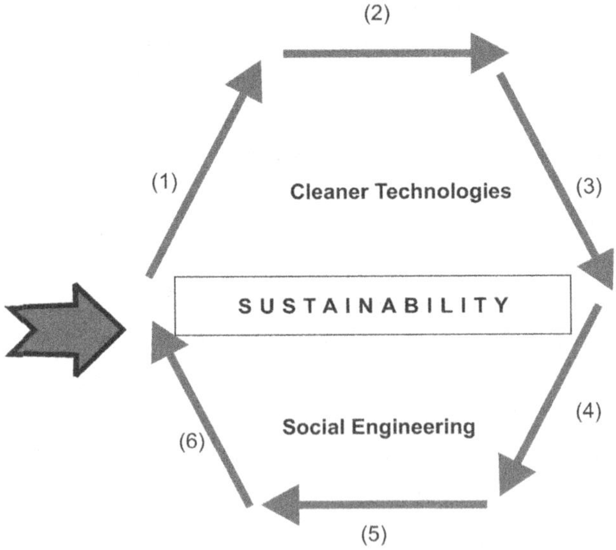

Fig. 11.1 *The model shows the six-step procedure to ensure sustainability*

Let us see the requirements one by one:

(1) Conserve use of all resources to the extent possible Generally, water, fuels, land, forests, minerals, chemicals, and other materials, are some of the resources required. Evidently, the advice is to adopt processes requiring least resources. How can this be done?

- Use more modern manufacturing processes which need less of these resources, like water.
- Use locally available materials for construction as far as possible. Avoid transport of materials over long distances.
- Use natural treatment processes (ponds, lagoons, wetlands, etc.) where land is available for the same. Natural treatment saves on fuel and does not need machinery.
- One would also use alternate sources of energy such as wind/solar/biogas operated systems as far as possible rather than draw electrical power from the town's supply system.
- Require permits for abstraction of ground waters. Give no free power. Find other ways for helping farmers.
- Ensure correct pricing of water (not by cross-subsidy) to make water conservation worthwhile.
- Use bio-fuels and other alternate fuels which give less carbon or no carbon and are more environment-friendly.

(2) Reduce waste, recycle and/or reuse Reduce both waste of materials in manufacture and waste of water and other resources. Reduce by applying the mantra of "Reduce, Reuse, Recycle" to all the manufacturing and treatment processes. In other words adopt "Cleaner Processes" as far as possible in both manufacture and waste treatment. Recycle or reuse after adequate treatment either in the production process or on land for irrigation. Thus giving zero discharge effectively.

(3) Minimize all health, social and environmental impacts For minimizing various health, social and environmental impacts there are various MOEF controls/guidelines for site selection, Environmental Impact Analysis (EIA), Life Cycle Analysis (LCA), ISO requirements, etc., which have to be respected. Industries have to be assured that besides protecting the environment overall costs will be reduced by adopting cleaner technologies.

Now because of Global Warming, we have to take care of gaseous emissions also, no matter where the site. Emissions of CO_2, CH_4, etc., can be avoided by process change, change of fuels, use of cleaner technologies, etc., and such other methods. Several methods are the subject of research at present.

Then come the Social Engineering steps.

(4) Promote awareness and education to people and staff to promote self-motivation It is necessary to promote awareness and education among the people, especially in the project staff, in order to involve the people and "motivate" them to do the right thing instead of monitoring them all the time and enforcing rules. Educate and involve people. Make them aware! Make them self-motivated. Have an "inclusive" program, leave out nobody, especially not the scheduled castes and tribes.

(5) Increase revenue through resource recovery *For municipal bodies*: resource recovery leading to increase in revenue is possible in many ways such as sale of biogas, conversion to electricity, sale of re-usable water, sludge, fibres, metals, food crops, algae, duckweed, fish, ducks, geese, etc.

To increase income from reuse of effluents via irrigation, promote cultivation of more cash-yielding crops on sewage farms. Cultivate Mentha, Citronella, Bio-fuels (Jetropha, etc.).

Upgrade treatment plants to enable produce effluent fit for reuse and thus add to the revenue as far as possible. Promoting reuse of the effluent is considered one of the best ways of ensuring sustainability. People are generally self-motivated to take care of the things they have to use themselves (TN State, Agro-industries wastes, etc., have successfully made reuse compulsory ("zero" discharge) in their State).

(6) Finally, in tune with the global ecosystem and human nature Use the revenue so generated to *internalize* benefits to the staff. Give it back to the staff or the Company, so as to generate self-motivation (not to the Government or Municipality). The staff will keep things running and thus ensure sustainability. Again, make sure the program is "inclusive", leaving out nobody.

In industries: Process selection, site selection, effluent reuse, "zero" discharge, etc., are all done willingly and creatively when benefit is internalized.

> Earning money from a treatment plant gives self-motivation to the staff for running it well; and self-motivation in turn gives sustainability.

> ***Example:*** It would be better if a wastewater treatment plant and irrigation farm operate as one unit, and income from sale of effluent for irrigation goes to treatment plant staff (not Government) as incentive for ensuring sustainability.
>
> In this way the wastewater treatment plant would run better and farming would also improve.

11.8 CASE STUDIES

A few illustrative cases are given below to facilitate discussion among participants.

Case study 1: A project for minimizing carbon emissions is developed in Pune. Is it likely to be sustainable?

A building in Pune (Orange County, near Pashan) has provided solar PV panels and wind turbines to generate a part of the electricity consumed by the building. This so-called "green" energy is available in limited quantity and supplied free to the flat-owners of the building. Any additional requirement of electricity is obtained automatically from a "grid" supply which has to be paid for by each flat-owner according to meter.

This project is likely to be sustainable because the flat-owners are themselves the beneficiaries of the project. The renewable energy project being a part of their ownership building, its maintenance is the responsibility of the flat-owners. The flat-owners have to keep all the systems in the building running. The project would not have been sustainable if a landlord was the owner and the tenants were sold the electricity.

The "green" energy system, however, slightly increases the capital cost of the building while it reduces the electric bill for the flat-owners. In the above case, the attitude of the flat-owners in the long run will decide whether more ownership buildings will come up with such systems or not. If maintenance becomes a daily headache, the flat-owners may prefer to opt for the conventional grid system. If maintenance is not a headache and life of the components is fairly long as expected (10–25 years for various components), the system has a good chance of use by other builders.

After all free electricity is a unique selling point for a builder to help selling flats, perhaps better than providing a gym or a swimming pool for the tenants, which also costs the builder something extra. In fact, it can be argued that a gym is not of much use to elderly people in the building whereas free electricity benefits all. Money saved by the flat-owners is not the only point. As more and more buildings provide such facilities, the whole country benefits by saving its resources of coal and oil and reducing its emissions to the atmosphere.

Case study 2: Are municipal wastewater treatment systems using *mechanized* methods of treatment sustainable for small and medium-sized towns in India?

Their sustainability is suspect! They are often not sustainable. The technology is not quite appropriate for a country located in a sunny, warm climate. In fact, a problem faced in installing *mechanized* wastewater treatment plants (e.g., Activated Sludge or Trickling Filters or the like) is always poor sustainability in a country like India. The plants either tend to use too much electricity which makes them unaffordable for the people, or the equipment tends to get out of order in India's hot and dusty environment, and needs expensive repairs, or the operation of the plant demands highly skilled attention which again makes it unmanageable by the relatively untrained staff employed by local authorities. Often, more than one of the above difficulties occurs at the same time, and the plant performs poorly or not at all (and becomes unsustainable).

The people who are supposed to "benefit" from the project actually find out as time goes that they are not able to afford the process and are not able to manage its operation properly. So the project is eventually operated indifferently and does not attain its original objectives. It remains unsustainable.

Case study 3: What methods of treatment are likely to be more sustainable for the many small and medium-sized towns in India?

"Natural" methods of wastewater treatment such as ponds, lagoons, wetlands, irrigation, etc., are generally more sustainable in the many small and medium-sized towns of India. The technology actually depends on natural organisms (bacteria, algae, plants, etc.) and is therefore more suitable for our warmer climate. Natural systems do not go out of order with usage and continue to operate almost indefinitely. Unlike mechanized systems, natural ones do not need electric power to operate. Natural systems depend on vegetation, algae, microbes, natural soil filtration, etc., for treatment and therefore function well in our warmer climates and have less chances of falling in disrepair even if not properly operated. They are inherently more sustainable.

However, natural methods are all relatively land-intensive and, therefore, they are generally not used for larger cities and metropolitan areas where land is difficult to find. They are more suitable for small and medium-sized towns where the flows are smaller and land is generally available. Moreover, when flows are small, they can often be finally disposed after treatment either by direct groundwater recharge or by land irrigation, both of which help to conserve the water underground for later reuse. Thus, to achieve this advantage, decentralization of the wastewater collection systems is highly recommended for small and medium-sized towns of India in the interest of sustainability. This is an urgent reform needed in a warm country like India in the interest of surer treatment and conservation of resources.

Moreover, in this manner, several usable by-products can be produced (e.g., edible fish and fowl, algae, reusable water, etc.) which can all bring some income. If this income is used to benefit the operating staff monetarily, the sustainability that comes from use of a natural system of treatment is doubly enhanced. The income from by-products, however small, should not be diverted or used by the local authority for other purposes.

Case study 4: How can one reduce the cost and increase the sustainability of a municipal wastewater treatment plant for a small and medium-sized town in India?

The present way of providing mechanized treatment followed by release of effluent to a water-course (river) for possible pick-up by downstream farmers (if they wish to do so) is both wasteful and unsustainable. Firstly, it requires the treatment plant to meet river discharge standards (which are higher than irrigation standards) and, secondly, the downstream farmers have to pay a cess to the Government if they want to pick up the treated effluent for watering their farms. Nothing from the cess payment goes to the plant operators. It all goes to the Government coffers. No wonders, the plants soon fail to give expected performance in treatment. The project fails in terms of "social engineering" and hence becomes unsustainable.

This situation is obviated if a "natural" method of treatment is adopted and the irrigation system is a dedicated part of the treatment plant where the effluent is directly used for irrigation and not released to a water course. The treatment is relatively low-cost, natural treatment and it has to meet irrigation standards. In this way small and medium-sized towns in India can afford the cost of treatment, the treatment meets the acceptance of all authorities and the process is manageable by the staff operating the plant. Sustainability is ensured!

Case study 5: How can sustainability of forests be ensured in areas where deforestation may, in fact, be attractive?

Every tree is assumed to absorb from the atmosphere about 1.0 tonne of CO_2 in its lifetime of 30–50 years. This CO_2 remains sequestered as wood as long as the tree (or wood products made from it) are not ignited. A large forest may consist of several thousands of trees. Thus, a forest can absorb many thousands of tonne of CO_2 in its lifetime and hold it sequestered for many years. Thus, to have forests cultivated and cared for specifically for reducing CO_2 from the atmosphere (called "dedicated" forests) one has to make it profitable for the landowner. This is where the "Carbon Credit" scheme comes in. It pays for creating new forests where none existed before. This payment ensures sustainability of the dedicated forest.

Another scheme for a similar purpose is the REDD scheme (Reducing Emissions from Deforestation and Degradation). Deforestation is often resorted to for creating land either for agriculture or habitation. Deforestation also occurs in course of time because the local people want twigs and logs of wood for use as domestic fuel for cooking, heating, etc. This trend can be checked by funding the land-owner to set up what are called "gobar-gas" plants under this scheme (see Section 7.2) so it becomes attractive to retain the forest.

India has found a similar way of achieving this objective by allowing the tribal people of the area or the local villagers to guard the forest produce and in return benefit from its sale. This has proved to be very effective wherever it has been done. An example of the benefit accruing to a tribal community can be given from Shahapur in Thane District or from Medha in Gadchiroli area, both in Maharashtra State, where a 1700 ha tract of forest land, denuded over the years, was beautifully re-forested after the local tribal people were given a stake in its development. A joint forest management committee entitled them to benefit from sale of products like gum, dried leaves, flowers, bamboo and mahua oil. The villagers were also entitled to 50% of the income earned by the forest department from the auction of timber. This transformed the deforested area into a re-forested area and generated employment as well.

Case study 6: Can we apply this approach to facilitate land acquisition in India?

Yes, we think it can be done. Land is a scarce resource in India. And much land is going to be required for setting up solar energy plants, wind farms, nuclear energy plants, hydro-power plants, etc. In fact, the old method of land acquisition through governmental channels has proved unworkable in many cases for two reasons. One reason is that the price offered is usually considered low and the local politician sees an opportunity in the situation. Secondly, the compensation received by the villager/agriculturist is generally soon frittered away and the poor villager ends up discovering that "he has sold the family silver" and is now even without any source of employment. Various alternatives have been tried to improve the situation.

Perhaps, as briefly mentioned in Section 1.4, the most logical approach seems to be to do the following: The local villager or the adivasi whose land is required for setting up an industry or a mining project or a public purpose infrastructure project (road, railway, municipal water or wastewater treatment plant, a power plant, etc.) should be firstly offered a residence somewhere else to take care of his displacement. Secondly, payment for his land should be made in two forms: Partly in cash, like a down-payment, and partly in kind, for long term benefit (such as a job in the industry or mine, a set of shares or bonds in the proposed venture which will fetch a regular income to the person in the future. It is important that the future remains protected. If this is done, he will part with his land. There is one more aspect, safety, which is extremely important in case of nuclear projects.

Case study 7: Increasing groundwater availability through rainwater harvesting increases the prosperity of villagers.

It is evident that if better methods of rainwater harvesting are used (better technology as explained in Section 3.2), more groundwater recharge will occur and the water table will rise. Water will become available to the local farmers for a longer period in the year, especially in the summer time, and thus change the cropping pattern. With more crops/fruits/flowers available for sale, better marketing arrangements will be set up. As the local people will be involved in the farming and marketing operations, sustainability will be ensured. One could hope that the face of India will change gradually.

At least two Indian villages can be held up as successful and sustainable examples of the above type (Section 6.4): Rallegaon Siddhi in Maharashtra (some 80 Km from Pune) under the guidance of Anna Hazare, and Vansda district in Gujarat under the guidance of Dr Kirtibhai Vaidya. Surely, many more such examples can be found.

11.9 CONCLUSION

From the above, we see that sustainability can be ensured in whatever we do (manufacturing, consulting, business or service), if resources are made to last longer and people are benefitted from supporting the activity. Sustainability will be automatically ensured. People's support has to be first ensured by passing on the financial benefit from the operation to them. The resources question is addressed through technology while the people question is addressed by finding a satisfactory answer to the eternal question: *"what's in it for me?"* These two basic principles must be satisfied for all environmental and infrastructure developmental activities to remain sustainable.

> The true value of a project never shows up on the balance sheet.

REFERENCES

Arceivala, S J, "Global warming and Climate change" Published by The Indian Environment Association, Nov 2010.

Arceivala, S J, "Green Technologies" Under publication by McGraw Hill, Noida.

The Orange County Foundation, Pune.

The Kirloskar Corporate Office, Pune (A green building) Published in Inside Outside Magazine, Feb 2010.

Energy Conservation Building Code (ECBC), Bureau of Energy Efficiency (Govt of India), R K Puram, New Delhi.

Additional Reading

Center for Environment Research and Education (CERE) Website, www.cere-India.org

The internet: The reader is advised to refer to the internet and select topic(s) of special interest.

EXERCISE QUESTIONS

1. In an effort to reduce the carbon footprint, what lifestyle changes would you like to suggest to the people of your city?
2. What fiscal incentives can you suggest for promoting greater use of "green buildings" and "green infrastructure"?
3. What steps would you recommend for greater adoption by industries of the following: Energy Audit, Water Audit, Waste Audit
4. Give examples of Corporate Social Responsibility and Corporate Social Irresponsibility?
5. Give more examples to show that sustainability can be improved by ensuring use of "cleaner technologies" along with "social engineering".

PROJECT/ASSIGNMENT TOPICS

Briefly discuss each of the following options for implementation, marking those which you think are feasible to adopt, and those which might possibly have any undesirable consequences.

Soft Options

- Hold seminars on green buildings, green infrastructure, energy and water audits.
- Devise incentives for green buildings, preferably second generation green buildings.
- Arrange energy audits and water audits in industries.
- Hold conferences on Corporate Social Responsibility matters.

Hard Options

- Implement incentive scheme for green buildings.
- Implement green infrastructure projects covering water, wastewater, solid wastes, etc.
- Implement mass transport projects in time bound programs.
- Implement energy and water audit reports in industries.
- Implement relevant CSR projects with industry funding.
- Implement "cleaner technologies" and "social engineering" programs to improve sustainability.

12 CHAPTER

Work Opportunities, Treatment Costs and Funding

12.1 Introduction

12.2 Opportunities in the Environmental Field

12.3 The New Build-Operate-Transfer (BOT) System

12.4 Manufacturing Business Opportunities

12.5 Employment Opportunities

12.6 Cost of Water and Waste Treatment

12.7 Funding of Projects

The environment is huge and there is enough for all. The trick lies in finding your "niche".

12.1 INTRODUCTION

The environmental field is a wide one. It is not the domain of only civil engineers. Various other engineers such as mechanical, electrical and chemical engineers and computer and software specialists are also required. Hence, this section was included so that everyone would know what variety of work is involved in the environmental field and can best judge for themselves their scope for success.

12.1.1 What are the Various Opportunities?

There are many opportunities for work in the environment field. Some of these opportunities are for civil engineers, some for environmental engineers and some for other engineers. Broadly speaking, opportunities exist for all in the following:

- ➢ Consulting services
- ➢ Specialist services
- ➢ Contracting business
- ➢ Turnkey. "Design-Build-Operate-Transfer"
- ➢ Manufacturing business
- ➢ Teaching and research
- ➢ Employment in public and private sectors

One can either be *employed* or one may become the *owner* and run any of the businesses.

As some may prefer to remain in employment while others may have a distinct preference for business, no two people can really be given the same advice. One has to decide for oneself. One can set up business readily today. But, business is not for all people. One has to have a business temperament.

12.1.2 What is "Business Temperament"?

It is not easy to describe it. But one can say in short that it consists of a few qualities such as the following which a successful businessman usually has. These qualities can also be developed as time goes:

- ➢ Man and money management ability

- Ability to articulate one's viewpoint and deal with clients gracefully
- Confidence in one's work
- And, lots of good luck!

At first entry level, it is always advisable to take up employment to get experience Civil/Environmental engineers can work with general civil or environmental engineering contractors preferably on *construction* works to gain experience during the initial period. Mechanical-electrical and chemical engineers can start with learning the *manufacturing* business and company's supply chain.

It is always necessary to learn man-management and money-management at somebody else's expense. Then, after some experience you may start one of the following businesses in environmental engineering and sciences.

12.2 OPPORTUNITIES IN THE ENVIRONMENTAL FIELD

12.2.1 General Consulting Services

Consulting services include general planning and design work in environmental engineering and sciences, EIA, Modelling, Detailed engineering, Tendering, Supervision during construction, start-up and commissioning.

The main advantage in starting a consulting service is that not much capital is required. For the same reason, a consultant generally has to face more competition because everybody (including retired officers) likes to call themselves as consultants. It is, therefore, better to find a "*niche*" within the field to practice in. Alternatively, build a "mix" of competent (intelligent) staff of engineers and scientists to be able to handle a variety of assignments. Be prepared to face competition from equipment suppliers, turnkey contractors, large specialized labs, and retired engineers who are all anxious to do the same kind of work.

12.2.2 Specialist Consulting Services

Specialist consulting services are rendered in a highly specialized or "niche" field selected from the ones noted below concerning the environment. Evidently, good prior experience is needed in the selected field before one can practice in it. Often, therefore, one joins a firm as an assistant and climbs up. There is scope for both engineers and scientists.

1. Lab analytical services. Field sampling/testing.
2. Environmental Impact Assessment; remote sensing; air/water monitoring.
3. Computerized modelling for air, water, etc., including design/ drafting.
4. Specialized software development and application.
5. Ground water exploration; tube-well sinking.
6. Plumbing in buildings for water and sanitation.
7. Air pollution monitoring and control (stack sampling, ambient monitoring, etc.).
8. Industrial heat and noise control.
9. Project Management services.
10. Sponsored studies; Contractual R&D work.
11. Hydrogeology/Geological Engineering for projects.
12. Foundation engineering /Structural engineering.
13. Oceanographic surveys and studies.
14. Risk assessment / Disaster management.
15. Corrosion control.
16. Drip irrigation projects.
17. No-dig technology applications.
18. CCTV surveys of sewers, etc.
19. Eco-tourism.
20. Waste audit and waste minimization in industry.
21. Control of gaseous emissions (Greenhouse gases) and fuel audit.

12.2.3 Contracting Services

Contracting services include construction work and therefore need much more capital and other resources. Also, scope for profits is higher. Hence, one needs a higher level of "manpower and money management capabilities". Business temperament is very essential.

The general stepwise progression in the contracting field is to get some experience first, then to undertake contracting work based on someone else's designs, then undertake contracting work with own designs, and finally work with one's own designs and own finance on "Build-Operate-Transfer" basis, as indicated below:

Experience → Gain contracting experience → Turnkey contracting → Turnkey contracting + finance (see below item 5 for examples BOT basis)

Obviously, the greater the risks involved, the greater the profits expected and the lesser the competition expected.

> In the old world, financial capital was the scarce resource. Talent is now the new value-driver. It is not just about money. It is about involvement and freedom of expression. Ownership per se isn't the issue; it's how you reward talented people for the fruits of their intellectual capital.
>
> Joel Kurtzman in "MBA in a Book"

12.2.4 Other Privatization Opportunities

Opportunities in this category start with rendering *"Contractual Services"* such as the following in which a specific job has to be done. For example:

1. Operation and maintenance of municipal wastewater treatment plants
2. Operation and maintenance of industrial wastewater treatment plants
3. Industrial hazardous wastes collection and transport from industries to disposal site
4. Industrial hazardous wastes treatment and disposal
5. Hospital wastes: collection, treatment and disposal
6. Municipal Solid Wastes: Collection and transport to disposal site
7. Gas collection from Municipal Solid Wastes Landfill sites
8. BPO – (Back-office Process Outsourcing)
9. Rainwater harvesting
10. Water supply meter reading and billing
11. Leakage detection and prevention
12. Financial management
13. Environmental Law

12.3 THE NEW BUILD-OPERATE-TRANSFER (BOT) SYSTEM

In reality, Build-Operate-Transfer (BOT) is a relatively new system in contracting in which the entire investment for a project comes from the contractor (or *his* Bank). The design and construction is done at the contractor's risk. The owner (*beneficiary*) only gives his land and the right to operate the plant for a contracted period (say 20–30 years) and assures the contractor to buy back the product (biogas or final water or reusable effluent) at an agreed price (*escalated* per year) during that period. At the end of the contract period, the contractor hands over the project to the beneficiary.

The contract period is important as the contractor earns back his investment + profit during this period. This new method is often preferred as the owner gets the project at no capital cost to himself. What is more important, he is assured of the quality of work because the contractor has to maintain the same for a long enough period of 20/30 years after construction, so he uses the best technology available to him at that time. Of course, the product (water, biogas, etc.) works out a little more expensive than if the investment was made by the owner himself since somebody has to bear the interest cost on the loan, and make some profit too.

As the contractor has to put up the money upfront, only contractors with deep pockets can venture into this game. Of course, as the contractor has to run the plant for the full duration, he tends to be more careful about the technology he chooses. Also, he can claim accelerated *depreciation rates* allowed by the Government during the period for tax purposes.

> The advantages of BOT and similar systems is that the ultimate owner makes no investment and is better assured of quality of workmanship.
> The contractor ensures good plant design and construction as he has to run it himself for a number of years after construction. The risk is entirely with the contractor. He negotiates with a Bank and other statuary authorities for permissions and is paid at an agreed rate only upon delivery of final product.

There is, however, a downside to this system which in some cases may lead to much litigation. As a long period of 20/30 years is involved, a lot of things can happen in the meanwhile. The beneficiary or his successor may in spite of the contract refuse to accept the product on some pretext or the other. An industry may go through a bad patch and may close down. The technology may change. In such cases, how does the contractor recover his costs if he is suddenly told before the expiry of the contract to stop supplying the product (reclaimed water or gas or whatever)? A contractor, on the other hand, may also cheat and produce more product/water or gas, etc., than contracted and sell the surplus to others making undue profit while using the original land and effluent free of cost.

A safer alternative sometimes preferred is to restrict the arrangement to "Design-Build" only. The contractor designs the plant and builds it on turn-key basis, while the beneficiary pays for the plant. Other variants depend on who pays and who owns the facility, the acronym may also be called BOO (Build-Own-Operate in perpetuity), or BOOT (Build-Own-Operate-Transfer after some years)

Whatever the basis, BOT, BOO, BOOT or design-build, the following examples are often quoted:

1. Complete water and wastewater treatment system for industries (e.g., Tirupur, TN).
2. Common effluent treatment plants (CETPs) for industries.
3. "Secure" landfill sites for hazardous industrial wastes.
4. Incineration plants for hazardous industrial wastes.
5. Water reclamation and reuse plants in industries (with cost recovery from water reused).
6. UASBs in industries to treat wastewaters – (cost recovery from biogas produced and sold back to the industry)
7. Municipal Solid Wastes – bio-methanation, gasification plants. (cost recovery from bio-gas sold).
8. Green buildings and Green infrastructure.

In many of these items, one can clearly see that all kinds of engineers are required (not just civil engineers). In fact, even lawyers and Chartered Accountants are required.

(In a lighter vein, Build-Operate-Transfer is what we do all the time for our children! We build them strong from birth, we operate their bank accounts when they are young, and finally we transfer our wealth to them when we die!)

12.4 MANUFACTURING BUSINESS OPPORTUNITIES

Mostly, mechanical, electrical and chemical engineers are needed in manufacturing business. In manufacturing a product, the work may be done either on direct ownership basis or on agency basis. Brand value is important to keep in mind in all cases.

1. Pipes, valves, meters, flow-measuring equipment.
2. Dosing equipment, mixing and aeration equipment.
3. Complete water and wastewater treatment plants (all units).

4. Solid wastes conveyance and disposal equipment
5. Air pollution monitoring, heat and noise control equipment
6. Manufacture of various chemicals used in water and wastewater treatment.
7. Manufacture of equipment for lab analysis
8. Manufacture of instrumentation and automation systems

One can list the various types of products that can be manufactured in this or any other business involved. One can add to the products as time goes. It is up to the individual and the financial backing and other support that one can get for entering a field. Building up "Brand Value" is something which happens with time – provided one consistently maintains one's quality and one's principles in day to day operation.

12.5 EMPLOYMENT OPPORTUNITIES

Employment is generally available with all Governmental and public sector organizations as well as private sector companies. Long-term employment is generally possible more with Governmental and public sector organizations such as the following:

- Government departments
- Water supply and sewerage agencies
- Pollution control authorities
- Public sector industries

Intelligent and "compliant" working is expected from all employees, especially from those who wish to have a long tenure of service.

Colleges, Universities, Research Institutes are also often in the public sector, but they generally offer some scope for adding to income (through consultation work and contractual R&D work). However, PhD is normally required for entry into this group.

Choose whatever you are happy with and be ready to take the plunge! GOOD LUCK !

> It is good to remember:
>
> BUSINESS HAS A CHARM,
>
> BUT EMPLOYMENT DOES NOT HARM !

12.6 COST OF WATER AND WASTE TREATMENT

Capital cost of a treatment plant whether it is for water treatment or wastewater treatment or industrial waste treatment, includes several different costs such as those noted below:

- Design and consultation cost
- Land cost, including legal fees
- Cost of survey and soil investigations
- Plant construction cost
- Equipment and erection cost incl electrification
- Testing and trials
- Interest charges on borrowings during construction period

The plant construction and equipment cost in turn depend on the nature of water or wastewater, its volume of flow, the degree of treatment required and its site conditions. It can thus be given only in approximate terms for a few cases as shown in Table 12.1.

Table 12.1 *Approximate capital cost for treatment*

Type of Wastewater	Capital Cost ₹ (lakh) per 1000 cu m daily flow
Municipal sewage	20–30
Cotton textiles (excluding denims)	50–80
Sugar	80–120
Pharmaceuticals	200–300
Soaps, detergents, vegetable oil refining	280–320
Bulk drugs, antibiotics, distilleries	500–900
Petrochemicals	800–1,200

* Municipal water supply costs about ₹10–18 per 1,000 L.

From the above table, it is apparent that a strong incentive exists with all types of industries for reducing the wastewater volume. As the volume is reduced, the total cost of treatment reduces.

Costs of municipal as well as hazardous solid wastes are difficult to indicate as many different variables are involved such as land costs, labour costs, transport costs, composting or dumping costs, manure and gas disposal costs, etc.

12.6.1 Annual Operation and Maintenance Costs

Annual operation and maintenance costs of a treatment plant are made up of the following costs:

- Direct operating costs such as staff costs, chemicals, fuel, electricity, transport, maintenance and repairs, insurance, etc.
- Fixed costs for amortisation and interest on borrowings.

These costs again vary widely depending on the nature of water/wastewater and its volume, the type of treatment given to meet standards and the terms on which funding has been obtained. Here again, strong incentive exists for reducing amount of waste to be treated.

Only an indicative idea can be given below of the annual O+M costs:

Source of Wastewater	Annual O and M Costs
Municipal	About 10% of capital costs (approx.)
Industrial	₹ 3–5 or more per cu m of effluent treated + cost of return of capital with interest
Hazardous wastes	Pretreatment costs + *secure* landfill costs (generally, both are higher than above)

12.6.2 Comparison of Costs of Different Treatment Methods

Different processes have different capital and operating costs. These two costs are not directly additive. Hence, either all annual costs have to be capitalized or all capital costs have to be annualized, assuming the simple rate of interest ruling at the time (say, 10% per year).

But all capital costs are not incurred in one year and operating costs escalate at different rates from year to year. Hence, comparison is better done using what is called the "*present worth basis*".

To find the present worth of a treatment project using two different processes or methods of treatment, it is necessary to know the following, or assign assumed values based on experience:

- Total capital cost of each treatment method to be compared.
- No of years for completion and likely capital expenditure in each year.
- O&M cost in first year and escalated O&M cost/year thereafter.
- Discounting factor or interest rate to convert later sums to net present value.
- Life of plant and salvage value at end.

By determining their present worth, one can compare two or more different methods of treatment and decide which method would be cheaper to adopt over the life of a plant.

Example 1: An extended aeration plant of 5 Ml/day capacity is expected to cost ₹ 21,876,000 and the land another ₹ 2,400,000. Construction will take two years. The operation and maintenance cost is estimated to be ₹ 5,071,000 per year in the first year after start up. Assume a 20-year horizon and the inflation rates as 9.5% in 1995–96, 8% in 1996–97, 6% in 1997–98 and each year thereafter. Using a discounting factor (i.e., interest rate) of 12% per year determine the present worth of all moneys which will be spent in constructing and operating the plant over 20 years. Neglect any salvage value.

The stepwise computation given in Table 12.2 helps to estimate the present worth of the Extended Aeration plant to be totally ₹ 76,535,500. Let us compare this plant to an Upflow Anaerobic Sludge Blanket (UASB) plant whose capital cost is more or less the same, but whose power consumption is only half as much. Using a similar calculation method, the present worth of the UASB plant is found to be only ₹ 49,984,000 instead of ₹ 76,535,500. Thus, the UASB would be cheaper in the long run although its capital cost is about the same as the extended aeration plant (Arceivala,1986). One would commit a folly by neglecting the operating costs and only deciding on the basis of capital costs.

Incidentally, one finds from the above computation that the total investment cost during the life span of a plant is typically two to three times higher than the initial capital investment cost of the plant. This is because of operating costs.

12.6.3 Some Planning Possibilities

The sample calculation shown above could also suggest to an entrepreneur that a natural (non-mechanized) process will have a lot of salvage value at the end of its life since land values in every country keep escalating from year to year. It would, therefore, be profitable to build a low cost "natural" system in the initial years and, later when the population grows one could switch over to a higher cost mechanized system, using a part of the profits made from sale of surplus land to meet the capital cost of the new mechanized system.

The advisability of building a new wastewater treatment plant on BOT basis should also be considered rather than waiting for long periods to secure "soft loans" from governmental or multinational sources. Industries are quicker to benefit from BOT type arrangements, when they can get usable bye-products like biogas or water from it.

Table 12.2 *Present worth of a 5 MLd extended aeration plant (₹ in lakhs)(Arceivala, 1986)*

Year	Capital Cost			Annual O & M Cost	Base Cost	Inflation Rate	Inflated Cost	Discounting Factor	Present Worth
	Civil	Mech. Elec.	Land						
1.	68.700	—	24.000	—	92.700	—	—	—	92.700
2.	103.060	47.000	—	—	150.060	1.180	177.071	1.250	141.657
3.	—	—	—	50.710	50.710	1.250	63.388	1.400	45.277
4.	—	—	—	50.710	50.710	1.330	67.444	1.570	42.958
5.	—	—	—	50.710	50.710	1.410	71.501	1.760	40.626
6.	—	—	—	50.710	50.710	1.480	75.051	1.970	38.097
7.	—	—	—	50.710	50.710	1.580	80.122	2.210	36.254
8.	—	—	—	50.710	50.710	1.680	85.193	2.480	34.352
9.	—	—	—	50.710	50.710	1.780	90.264	2.770	32.586
10.	—	—	—	50.710	50.710	1.880	95.335	3.110	30.654
11.	—	—	—	50.710	50.710	2.000	101.420	3.480	29.144
12.	—	—	—	50.710	50.710	2.120	107.505	3.900	27.565
13.	—	—	—	50.710	50.710	2.240	113.590	4.360	26.053
14.	—	—	—	50.710	50.710	2.383	120.841	4.899	24.667
15.	—	—	—	50.710	50.710	2.535	128.554	5.504	23.355
16.	—	—	—	50.710	50.710	2.697	136.760	6.185	22.113
17.	—	—	—	50.710	50.710	2.869	145.489	6.949	20.936
18.	—	—	—	50.710	50.710	3.052	154.776	7.808	19.823
19.	—	—	—	50.710	50.710	3.247	164.655	8.773	18.768
20.	—	—	—	50.710	50.710	3.454	175.165	9.857	17.770
							Total		**765.355**
			i.e. (present worth of money which will be spent over 20 years)						

12.7 FUNDING OF PROJECTS

Funding is generally available for undertaking infrastructure projects including construction of treatment plants, either as:

- Grants
- Repayable Loans
- Other forms

12.7.1 Grants

Grants are available (to a limited extent) from Government and public bodies for infrastructure projects. Grants may also be available (to a limited extent) from bilateral sources for infrastructure projects.

Grants become the preferred method of funding as generally no repayment is needed. Everybody likes free money! But, that is precisely the reason why grants are relatively scarce and only given to public bodies.

12.7.2 Repayable Loans

Repayable loans are given to both public and private bodies.

For public bodies, repayable loans are available from multi-lateral sources such as the World Bank, Asian and other Regional Banks, Bi-lateral agencies, national banks, governments, public bodies, etc. This is the common form of raising funds for infrastructure projects.

"Soft loans" @ low interest rates and/or longer repayment duration are sometimes given for special purposes such as infrastructure projects.

For industries and private organizations, harder terms may be specified. For industries and private organizations, repayable loans are mostly available from national banks. Capacity for repayment with interest charges is first evaluated. Some down payment from the beneficiary is also required.

12.7.3 Estimation of EMI

If an amount P is borrowed at interest rate (i) to be returned in (n) number of years, the amount to be returned per month (EMI) equals:

$$\text{EMI} = \frac{P \cdot (i) \cdot (1+i)^n}{12(1+i)^n - 1}$$

On this basis the EMI is calculated to ensure that the loan is returned to the funding agency with interest by the end of the payment period.

12.7.4 BOT and Other Such Basis

Another way of finding funds for providing infrastructure is to adopt the Build-Operate-Transfer (BOT) method so that the contractor obtains the necessary funding from his own sources or a national bank. The funding agency may ultimately still be the bank. However, the applicant is the contractor who takes the risk. The BOT system has already been explained in Section 12.3 above.

Human Nature!

Everybody likes to be in the clean water business. Few worry about wastewater whose treatment is neglected unless forced to do. Thus, our wastewater coverage ratio is very low compared to water supply. With such a scenario, responsibility for water and wastewater have to be given to one single authority and at a price enough to undertake both activities.

Wastewater treatment in Indian towns and cities has to be increasingly done by using "natural" treatment methods suitable for warm climates like ours, and cheaper to operate. Small industries can benefit from out-sourcing to common effluent treatment plants (CETPs). Larger industries as well as bigger cities may go in for mechanized systems which require skilled operation and maintenance.

All can benefit from adopting newer funding methods like "Build-Operate-Transfer" but legal agreements have to be drawn up very carefully to ensure that neither party takes undue advantage of the other.

Index

A

Acid rain, 153
Air pollution control strategies, 146
Algae, 5
ALMs, 278
Alternative sources of energy, 52
  Already in use, 53
  In development stage, 73
  Ready for use, 58
Ambient and indoor air quality, 131

B

Bacteria, 4
Bio-fuels, 71, 72
Bio-methanation, 240
Bottled water, 166
Build-Operate-Transfer (BOT) system, 309
Business opportunities, 311

C

Capacity building and community empowerment, 38
Carbon cycle, 11
Chains 9
Coastal Regulation Zone (CRZ), 200
Common Effluent Treatment Plants (CETPs), 203
Comparison of different treatment costs by present worth, 314
Composting, 242–244
Conservation, 27
Corporate and community initiatives, 274
Corporate Social Responsibility (CSR), 275
Criteria, 22

D

Desalination, 173, 224
Due diligence studies, 119

E

Earth's atmosphere, 132
Electrical and E-wastes, 260
Ecosystems, 1–13
Emission factors, 45
Employment opportunities, 312
Energy cycle 12

Energy (see alternative sources of energy), 52
Environment protection, 8
Environmental impact assessment (EIA), 119
Environmental management, 107
Evaporation losses from Indian lakes, 225

F

Fossil fuels (coal, oil, petroleum), 44
Funding of projects, 317

G

Geo-thermal energy, 75
Green buildings and area development, 279
Green buildings, second generation, 286
Green House Gases (GHG), 8
Green industries, 291
Green infrastructure, 288
Green transport, 287
Greening efforts, 279

H

Hazardous substances in E-wastes, 264
Hazardous waste disposal, overview, 246
Hazardous wastes monitoring, 254
Hazardous wastes process changes, 254
Hazardous wastes sites remediation technologies, 253
Hazardous wastes treatment, storage and disposal facilities, 250
Health hazards in irrigation, 213
Hospital wastes, 256
Hybrid systems, 69

I

Indian standards, 34
Indoor air pollution and ventilation, 150
Industrial stack emissions and control, 145

L

Landfills, 238
Life cycle assessment (LCA), 125
Linking of rivers in India, 30
Lysimeter tests, 217

M

Mangroves and wetlands, 201
Manufacturing opportunities, 311
Marine pollution, 197
Matrix for EIA studies, 124
Meteorology and air movements, 142
Modern trends in Environmental management, 112
Municipal solid wastes disposal, 55

N

Natural and mechanized technologies, 188
Natural resources and need for conservation, 27
Noise pollution, 154
Nutrient cycle, 12

O

Opportunities in Environmental field, 307–309

Opportunities in Business, 311
Ozone depletion, 152

P

Permissible air quality, 140
Policies and strategies, 18
Pollutants, accumulative, 8
Pollutants, persistant, 8
Power generation, 44
Preventive environmental management, 107

R

Rainwater harvesting, 79
  In rural areas, 89
  In urban areas, 80
Rainwater harvesting approaches involved, 95
Rainwater harvesting hazards, 102
Resource recovery (see waste-to-energy), 189
Reuse for augmenting public water supplies, 222
Reuse in irrigation, 213
Reuse of wastewater in India, 212

S

Sanitary landfills, 235
Secondary pollutants, 138
Secure landfills, 252
Segregation and ragpicking, 233
Site selection for new developments, 117
Slum sanitation problem, 193
Soil-crop-water balance, 213
Solar energy systems, 62
Solid wastes disposal by incineration, palletisation, etc, 239
Solid wastes disposal, recommended methods, 234
Solid wastes dumping in low-lying land, 237
Species diversity, 10
Standards, 22
Statistics, 38
Sustainability, 292

T

Thermal pollution, 196
Total sanitation in India, 183
Town-planning and infrastructure development, 116
Traffic air pollution and control, 147
Treatment cost comparison by present worth, 314
Treatment costs, water and wastewater, 313
Treatment for domestic wastewater, 190

V

Ventilation, 150
Vermi-compost, 240, 243

W

Waste disposal, overview, 230
Waste-to-energy plants, 189
Wastewater treatment for pollution control, 179
Wastewater treatment objectives, 182
Wastewater treatment technologies, 184
Water quality checks in distribution systems, 165

Water quality for general consumption, 162
Water quality for industries and boilers, 166,
Water resources development – the Indian scenario, 167
Water resources development projects, 122
Water sources, 158
Water treatment, 163
Wave energy, 74
Webs, 10
Wetlands, 201, 202
Wind energy systems, 59

Z

Zero discharge from industries, 194

Authors' Profiles

Soli J Arceivala pioneered the business of environmental consultancy in India in 1961, long before any legislation for pollution control existed. He founded the first consultancy company in India, called AIC, specializing in water and environmental engineering work. In 1993, the Company became AIC Watson, then Montgomery Watson and presently, MWH, with Prof Arceivala as Chairman and then Chairman Emeritus until 2001 when he retired. He holds a degree from Harvard University, USA, and is a member of several professional bodies.

Professor Arceivala was the Regional Adviser for Environmental Health with the UN / WHO Regional Office for Europe, and was later the Chief for WHO South East Asia Region, covering 11 countries. Before joining WHO, he was the Director of the National Environmental Engineering Research Institute (NEERI), Nagpur, Government of India, and before that Vice-Principal, VJTI, Mumbai.

He is involved in teaching, research and consultancy work. He pioneered water conservation and reuse in various Indian industries in the mid-sixties and several plants for reuse of water have been built in India and the Middle East based on his designs. In recent years, he has worked on developing reuse of water for augmenting water resources for community uses and exploring low-carbon technologies for waste treatment.

Professor Arceivala is the author of seven books on environmental engineering, three published by McGraw Hill, of which one has been translated into Turkish and Iranian languages. His other books have been published by the Indian Water Works Association, the Indian Environmental Association and Marcel Dekker, New York.

Every year, for 20 years, he has been invited to give lectures at IHE, Delft, Holland, on wastewater treatment. In 1993–1994, Prof Arceivala was the President of the Indian Water Works Association and in 1996 he was the President of the Indian Environmental Association. In 1998, he became the first Indian to be awarded "Distinguished Membership" of the American Society of Civil Engineers, (ASCE), USA, in honour of his lifetime work in India and overseas. Several Indian organizations have also honoured him with awards.

Dr. Shyam R. Asolekar is a Professor at the Centre for Environmental Science and Engineering at the Indian Institute of Technology, Bombay. He also served as the *Head of the Department* during May 2006 and June 2009. Dr Asolekar holds two degrees in chemical engineering: an undergraduate degree from Mumbai University's UICT (1980) and a master's degree from Indian Institute of Science, Bangalore (1985). Subsequently, he went to the U.S. to obtain his M.S. and Ph.D. in Environmental Engineering from Syracuse University (1987) and University of Iowa (1991), respectively. After pursuing his postdoctoral studies at the Harvard University in Boston he returned to India in January 1993 to teach at the IIT Bombay.

Dr Asolekar has been a *member* of the "Dahanu Taluka Environmental Protection Authority" (since 1997) as well as the "Expert Committee on Conversion of Municipal Solid Wastes to Energy" (since July 2005), both constituted by the Honorable Supreme Court of India. Recently, the Honorable Planning Commission of India has recruited him in the "Steering Committee on Environment, Forests & Wildlife" for the 12th Five Year Plan (2012–2017). He was in the "Working Group on Rivers, Lakes and Aquifers" for the 11th Five Year Plan as well as the core group for mid-term assessment of plan-period 2007–2012. Recently, Dr Asolekar has been appointed as a core-team member for a project envisaged and conducted by the consortium of seven IITs entrusted with the responsibility of preparing Ganga River Basin Environment Management Plan (GRB-EMP) by the Ministry of Environment and Forests (MoEF), GOI, New Delhi.

The Govt of Maharashtra had appointed Prof Asolekar from 2007–2010 as an expert member in the "expert committee" entrusted with recommending clearance for the *High-rise Buildings* after investigating environmental suitability of the tower projects having more than 20 floors or taller than 70 meter of height. Earlier, under the Chairmanship of Dr Madhav Chitale, the Govt of Maharashtra had appointed Dr Asolekar in the "Fact-finding Committee" to probe the causes of the collapse of civic infrastructure in the aftermath of the deluge in Mumbai between July 26 and 29, 2005. Recently, Dr Asolekar has been inducted by the MoEF into the technical expert panel responsible for developing and implementing strategies for management of toxic and hazardous waste dump sites in India.

His areas of research and industrial advice have been Treatment, *Recycle and Reuse of Industrial and Municipal Wastewaters, Rejuvenation of Lakes, Rivers, and Polluted Coasts, Development of Technologies for Sustainable Futures, Environmental Policy, Preventive Environmental Management, Eco-industrial Networking, Global Environmental Issues, Treatment of Hazardous, Biomedical, and Solid Wastes, and Application of Remotely Sensed Data for Monitoring of Environmental Systems.*

Made in United States
Orlando, FL
04 November 2026

90202239R00193